Finite Element Structural Analysis: New Concepts

Finite Element Structural Analysis: New Concepts

J. S. Przemieniecki
Air Force Institute of Technology (Retired)
Wright–Patterson Air Force Base, Ohio

EDUCATION SERIES
Joseph A. Schetz
Series Editor-in-Chief
Virginia Polytechnic Institute and State University
Blacksburg, Virginia

Published by
American Institute of Aeronautics and Astronautics, Inc.
1801 Alexander Bell Drive, Reston, VA 20191

American Institute of Aeronautics and Astronautics, Inc., Reston, Virginia

1 2 3 4 5

Library of Congress Cataloging-in-Publication Data

Przemieniecki, J. S.
  Finite element structural analysis : new concepts / J.S. Przemieniecki.
     p. cm. -- (Education series)
  Includes bibliographical references and index.
  ISBN 978-1-56347-997-7 (print : alk. paper) -- ISBN 978-1-56347-998-4 (e-book)
  1. Finite element method. 2. Structural engineering--Mathematics. I. Title.

  TA347.F5P783 2009
  624.1'710151825--dc22

                                                                    2009014064

Dedicated to my wife, Fiona

Foreword

Finite Element Structural Analysis: New Concepts by J. S. Przemieniecki introduces a novel concept for accurate assessment of solution accuracy of the powerful and versatile finite element structural simulation technology. This has the potential of significant impact in the development of related new analysis codes and also modification of existing ones, which will generate reliable solution bounds that are of great importance in the analysis of complex, practical engineering problems. Further, the textbook will also act as a rich source of reference material for researchers and practicing engineers. The author, Dr. Przemieniecki, is the leading authority in the field of finite element structural analysis, particularly related to aerospace engineering, and has been involved in associated fundamental development since its very inception. This text is a fitting successor to the author's earlier pioneering textbook, *Theory of Matrix Structural Analysis* published in 1968, which is the first comprehensive textbook related to aerospace engineering that had the most significant impact on the development of this technology and continues to be a highly referenced source of material both in teaching and practice. This text covers finite element developments pertaining to 1) assumed displacements distribution, 2) corrective distribution superimposed on assumed displacement distribution, and 3) assumed stress distribution within an element. Such a development covers a wide array of elements including plane plate (triangular and quadrilateral), solid elements (tetrahedron, prismatic pentahedron, and rectangular hexahedron). These developments are provided in a clear and detailed fashion, in which each element stiffness, thermal, and mass matrices are derived in careful detail. The author also provides a fascinating depiction of relative efficacy of the three techniques in the book, which also includes numerical data enabling a clear comparison of the solution results. The textbook is highly recommended for graduate- and senior-level undergraduate courses in engineering mechanics, particularly related to aeronautics and associated disciplines. The text will also provide invaluable guidance for practicing engineers in accurate modeling and simulation of problems encountered in industry.

K. K. Gupta
NASA Dryden Flight Research Center

Table of Contents

Preface

The motivation for writing this book arose from the fact that the existing methods of finite element structural analysis do not provide any means of assessing the accuracy of results obtained for any given mesh size relative to the size of the structure and distribution of the applied loading. The new method of analysis described in this text provides for a quantitative assessment of the accuracy. The proposed method consists of two separate sets of analysis.

The first analysis is based on traditional element properties derived from assumed displacement fields (some of which even violate the required equations of stress equilibrium within the element). Such elements are described here as displacement elements. These elements are designated in this text as T1, T3, T5, T7, T9, and T11.

The second analysis uses three types of elements: stress elements (T4, T6), enhanced displacement elements (T2, T8, T10), and a special element (T12). Stress elements are based on assumed stress fields satisfying the necessary equations of stress equilibrium. Enhanced displacement elements are based on assumed displacement fields with additional fields vanishing on the element boundaries, the magnitudes of which are determined from the Principle of Minimum Total Potential Energy. The special hexahedron element T12 is constructed from two pentahedrons based on assumed displacement fields.

The results from the first and second analysis are to be compared. If the results between the two analyses are judged to be close enough, the solution can be considered bounded. If not, the process is repeated with a smaller grid until a reasonable closeness of the two solutions is obtained.

The author is very grateful for the encouragement and support he received from Kajal K. Gupta of NASA Dryden Flight Research Center. He also wishes to thank Maj Mirmirani, dean of the College of Engineering, Embry-Riddle University, Florida, for preparing computer codes supporting the new analysis. Special thanks are also extended to Anita and Kevin Harper for their support in preparing the final manuscript for publication.

J. S. Przemieniecki
May 2009

Part 1
Introduction

1
Finite Element Analysis: An Overview

1.1 Introduction

The finite element method (FEM) for the analysis and design of structures and mechanical components is based on the concept of replacing the actual continuous structure by a mathematical model made up of elements of finite size (referred to as finite elements) having known elastic and inertia properties that can be expressed in matrix form. For this reason, early finite element methods were described as matrix methods of analysis, but today FEM has become the accepted terminology. The matrices representing these properties are considered as building blocks, which, when assembled together according to a set of rules derived from the theory of elasticity, provide the static and dynamic properties of the actual system. To put FEM in the proper perspective, it is important to emphasize the relationship between finite element methods and classical methods used in the theory of deformations in continuous media. In classical theory, the deformational behavior is considered on a macroscopic scale without regard to size and shape of the elements within the prescribed boundary of the structure. In finite element methods, elements are of finite size and have a specified shape. These elements are specified arbitrarily in the process of defining the mathematical model of the continuous structure. The properties of each element are calculated using the theory of continuous elastic media, and the analysis of the entire structure is carried out for the assembly of the individual elements. When the size of the elements is decreased, the deformational behavior of the mathematical model converges to that of the continuous structure.

Structural complexity can be illustrated for the case of the Anglo–French Concorde, the first supersonic passenger aircraft used commercially. A drawing of this aircraft is shown in Fig. 1. The author participated in the design and analysis of a precursor aircraft (Bristol Type 198) before it became the Anglo–French project. The Type 198 was a 130-passenger transatlantic supersonic aircraft similar to the later Concorde. The analysis performed by the author was based on a simplified finite element model of the wing/body structure shown in Fig. 2. This analysis used the force method (an early analysis bases on element forces) with about 120 unknown internal forces in the wing/fuselage structure. Today finite element models can use thousands of elements. An example finite element model of a modern aircraft structure is shown in Fig. 3. Another example is shown in Fig. 4 of the Boeing 777 aircraft, illustrating one-half of the aircraft. This example analysis used about 70,000 nodes and 175,000 elements, with 275,000 degrees of freedom. An example finite element model of an ocean tanker from the ship building industry

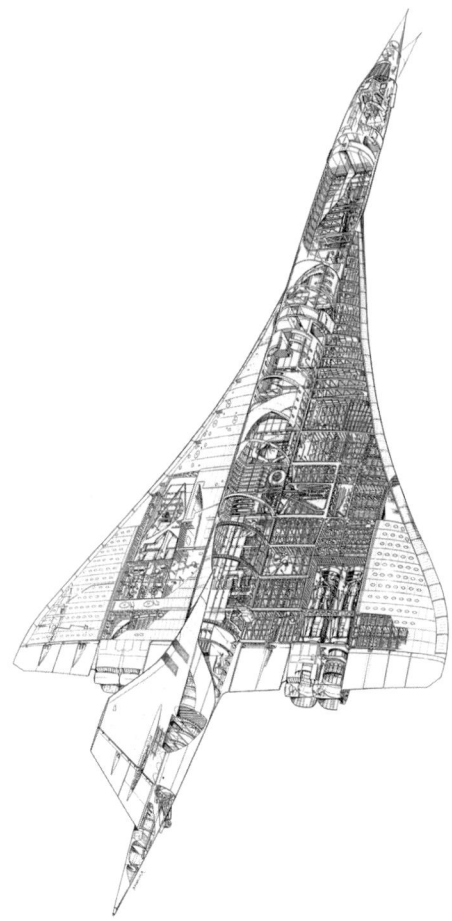

Fig. 1   Cutaway drawing of the Anglo–French Concorde (reproduced with permission of The Flight Collection, United Kingdom).

is shown in Fig. 5. Other applications include design of buildings, bridges, and mechanical components.

One of the most important steps in finite element analysis is the selection of nodal points and the size of finite elements representing the structure to be analyzed. Traditionally, the process is repeated with smaller grids to ensure adequate accuracy of the stresses and deformations; however, this process is computationally not very efficient and in some cases can produce inaccurate results if the grid-refining process is discontinued too early. An example of this problem is shown in Fig. 6, where the grid-refining process could be discontinued with 25 nodes because the analysis indicated only a small change in the resulting stress from 15 nodes to 25, but if the node-refining process were continued for another step with 45 nodes, then the resulting stress would have indicated a significant stress increase above the calculated stress in the previous refinement step. The proposed method of the

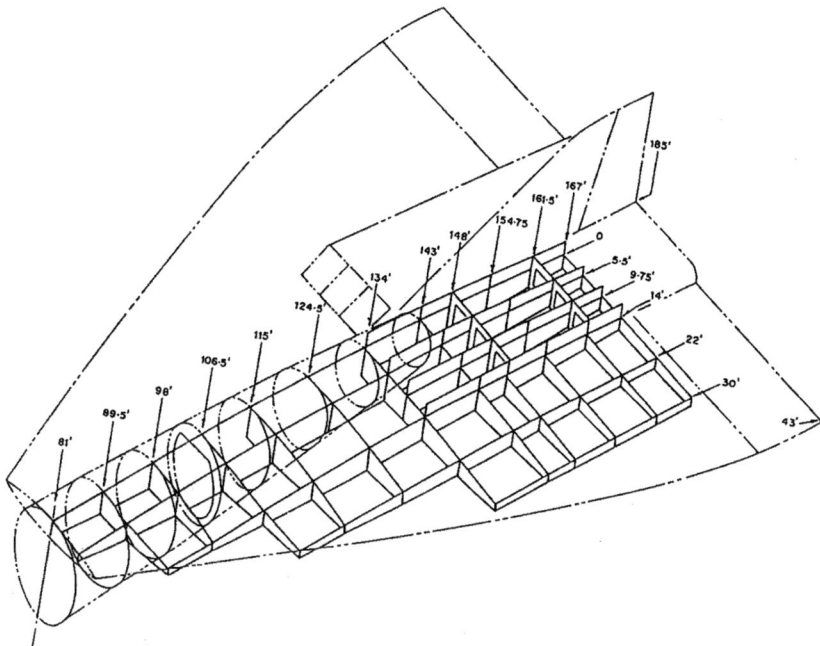

Fig. 2   Finite element model of the Bristol Type 198 aircraft, before it became the Anglo–French Concorde.

upper- and lower-bound analysis described in this text avoids this problem, and it provides a better measure of the expected accuracy for a given grid size.

Some typical finite elements used in structural analysis are shown in Fig. 7, where the bullet points represent element nodes at which displacements and forces are specified. Both two-dimensional and three-dimensional solid elements are shown. In all elements, for each node displacement there is a corresponding element force. Pin-jointed elements and beam elements are not included in this study because their properties are exact, and therefore they do not have lower- or upper-bound solutions.

1.2   Finite Element Analysis Procedure

Finite element analysis (FEA) consists of a sequence of steps that are illustrated here using a cantilever beam example represented by 12 elements (rectangular plates of constant thickness) shown in Fig. 8. These steps are typical for any finite element analysis.

1.2.1   Step 1: Formulation of the Element Grid and Boundary Conditions (Finite Element Mesh)

An example of grid lines and nodes is shown in Fig. 8. In this example there are 20 nodes (1–20) with horizontal and vertical displacements at each node and

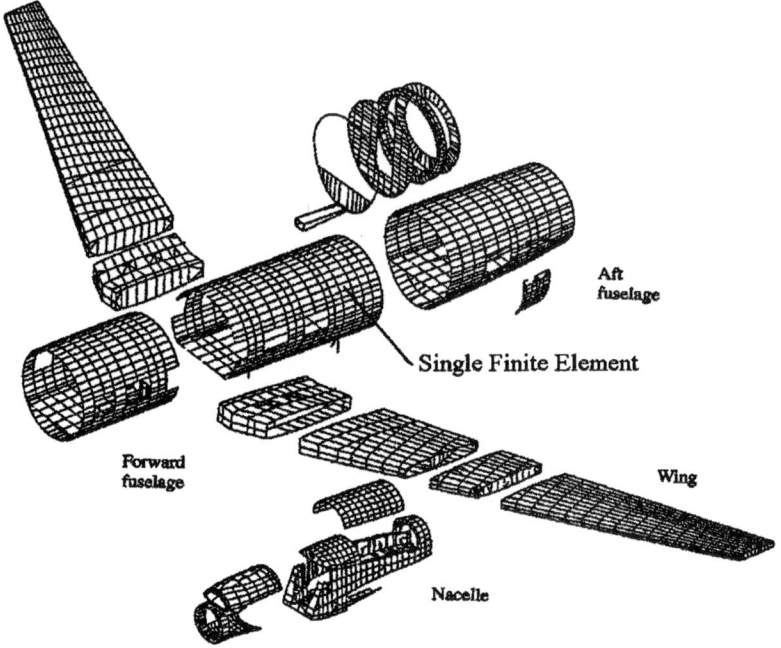

Fig. 3   Finite element model of a large aircraft (courtesy of Lockheed Martin Corporation).

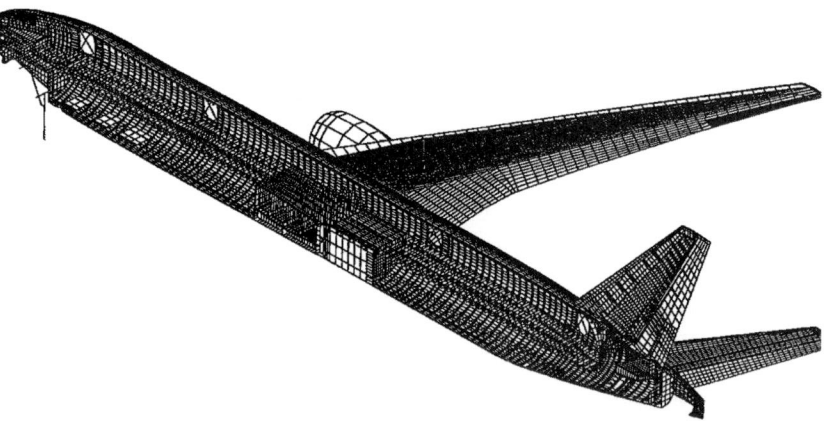

Fig. 4   Finite element model of the Boeing 777 aircraft (only one half of the aircraft structure is shown) (courtesy of The Boeing Company).

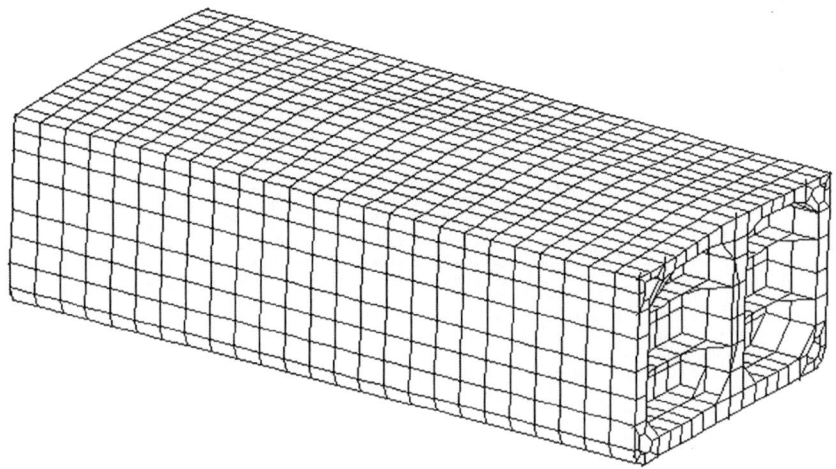

Fig. 5 Finite element model of a tanker (courtesy of American Bureau of Shipping).

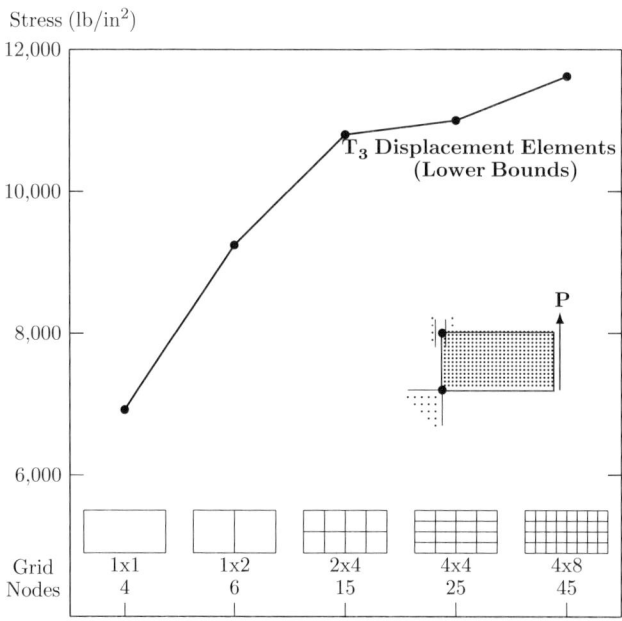

Fig. 6 Stress at the lower-left corner in a cantilever under distributed load P for various grid sizes calculated with the displacement elements: $P = 1000$ lb, $E = 10 \times 10^6$ lb/in.^{2}, $v = 0.3$, beam length = 24 in., beam height = 12 in., and thickness = 0.1 in.

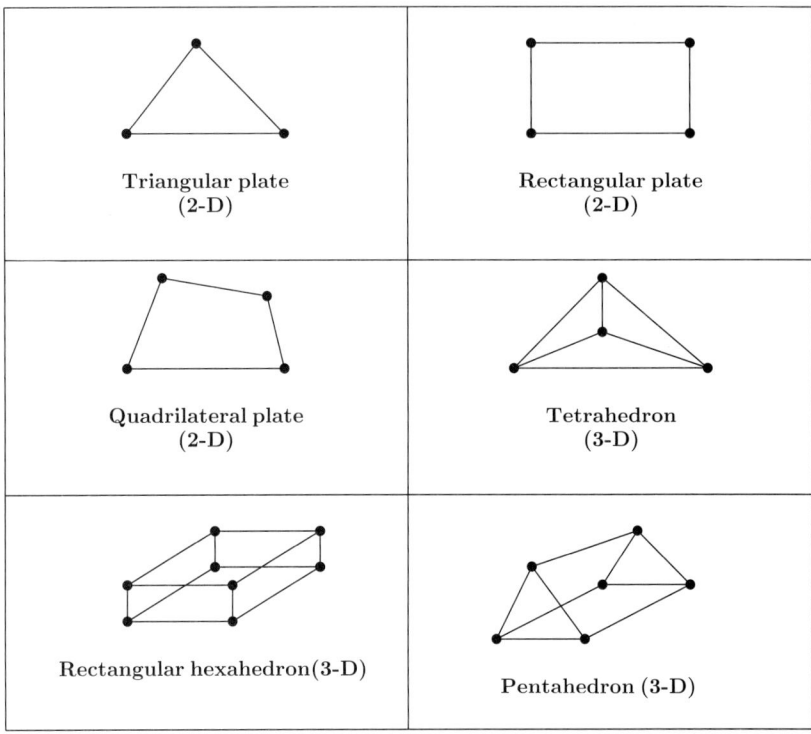

Fig. 7   Typical finite elements used in upper- and lower-bound analysis.

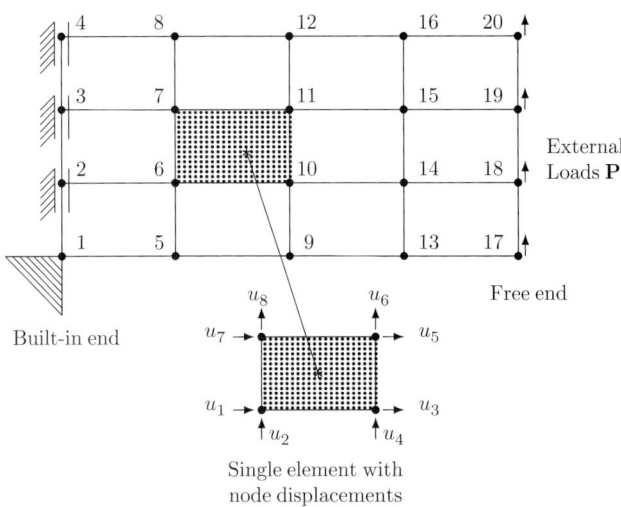

Fig. 8   Example of FEA: cantilever beam with transverse loads at the tip.

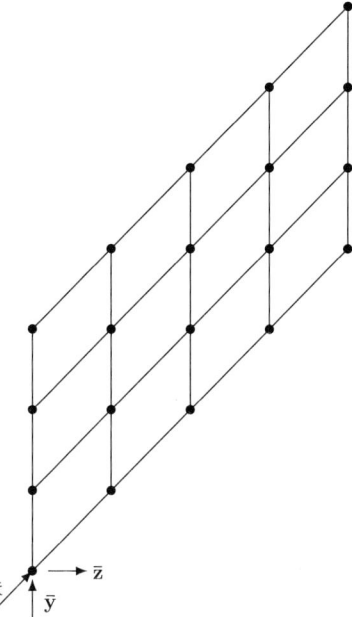

Fig. 9   Global coordinate system used for the two-dimensional structure in Fig. 8.

applied loads at nodes 17, 18, 19, and 20. At node 1, both horizontal and vertical displacements are zero. In addition, horizontal node displacements at nodes 2, 3, and 4 are zero to represent the "built-in" condition.

1.2.2   Step 2: Selection of the Reference Origin for the Global Coordinate System

The reference origin is used for the coordinates of the element nodes and applied loads or constraints on the structure. For the example in Fig. 8, the reference origin is at node 1. Two examples of the selected origin for the global coordinates are also shown in Figs. 9 and 10. In the example of Fig. 8, there are 20 nodes (1 ... 20) and 12 elements (1 ... 12). At each node (i) there are only two displacements: u_x in the x direction and u_y in the y direction. In a three-dimensional structure made up of plate elements, each node will have three displacements. For two-dimensional plate structures only two displacements at each node are needed.

1.2.3   Step 3: Selection of a Rigid Frame of Reference

A rigid frame of reference is needed to calculate the node displacements. For the example in Fig. 8, this is accomplished by defining displacements u_x and v_y at node 1 and u_x at nodes 2, 3, and 4 as zero. This establishes a rigid frame of reference with respect to which all other node displacements are calculated. This means that rigid-body translations and rotations of the whole structure are

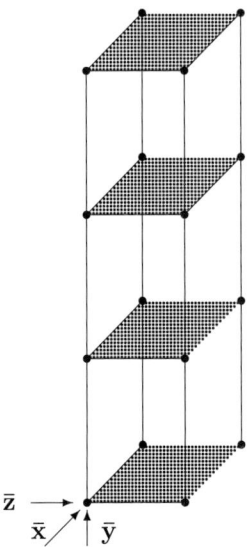

Fig. 10   Global coordinate system for a three-dimensional structure made up from plate elements.

eliminated. This is accomplished simply by eliminating all rows and columns in the assembled stiffness, thermal stiffness, and/or mass matrices corresponding to zero displacements defining the rigid frame of reference.

1.2.4   Step 4: Determination of the Direction Cosines for Each Element

The direction cosines are required to calculate element properties in global coordinates from the element properties in local coordinates. The direction cosines are the cosines of angles between the nodal vectors representing node displacements or rotations and coordinate vectors of the global frame of reference. The direction cosines for each element on the structure are assembled into a matrix denoted by λ. The method of calculating λ is discussed in Sec. 1.3.

1.2.5   Step 5: Computation of Element Properties in Local Coordinates—Stiffness k, Mass m, Thermal Stiffness h, and Thermal Load q

For each separate element the fundamental equation for the element properties is of the form,

$$\mathbf{S} = \mathbf{k}\mathbf{u} + \mathbf{h}\alpha T = \mathbf{k}\mathbf{u} + \mathbf{q} \qquad (1)$$

illustrated for a rectangular flat plate shown in Fig. 11.

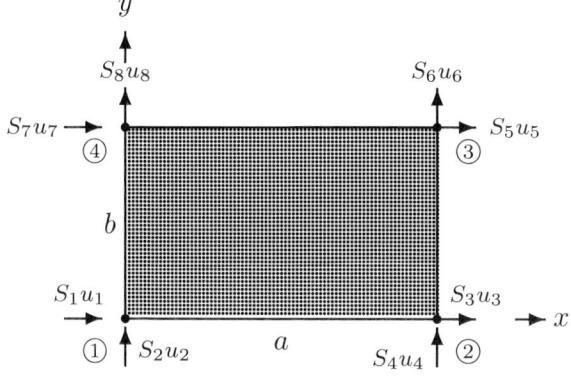

Fig. 11 Displacements and forces in a rectangular plate element.

The element forces **S** and element displacements **u** are given by

$$S = ku + h\alpha T = ku + q \tag{2}$$

$$S = \{S_1 \ S_2 \ \cdots \ S_8\} \tag{3}$$

and

$$u = \{u_1 \ u_2 \ \cdots \ u_8\} \tag{4}$$

where the element forces and displacements are expressed as column (6×1) matrices. (Curly brackets are used for column matrices to save space.)

The stiffness matrix **k** is given by

$$k = \frac{Et}{12(1 - v^2)} \begin{bmatrix} k_{1,1} & k_{1,2} & k_{1,3} & k_{1,4} & k_{1,5} & k_{1,6} & k_{1,7} & k_{1,8} \\ k_{2,1} & k_{2,2} & k_{2,3} & k_{2,4} & k_{2,5} & k_{2,6} & k_{2,7} & k_{2,8} \\ k_{3,1} & k_{3,2} & k_{3,3} & k_{3,4} & k_{3,5} & k_{3,6} & k_{3,7} & k_{3,8} \\ k_{4,1} & k_{4,2} & k_{4,3} & k_{4,4} & k_{4,5} & k_{4,6} & k_{4,7} & k_{4,8} \\ k_{5,1} & k_{5,2} & k_{5,3} & k_{5,4} & k_{5,5} & k_{5,6} & k_{5,7} & k_{5,8} \\ k_{6,1} & k_{6,2} & k_{6,3} & k_{6,4} & k_{6,5} & k_{6,6} & k_{6,7} & k_{6,8} \\ k_{7,1} & k_{7,2} & k_{7,3} & k_{7,4} & k_{7,5} & k_{7,6} & k_{7,7} & k_{7,8} \\ k_{8,1} & k_{8,2} & k_{8,3} & k_{8,4} & k_{8,5} & k_{8,6} & k_{8,7} & k_{8,8} \end{bmatrix} \tag{5}$$

where E is the Young's modulus of the material, v is the Poisson's ratio, t is the plate thickness, and $k_{i,j}$ are the stiffness matrix coefficients, which are shown in the chapter on element properties.

The element thermal stiffness matrix **h** is given by the column matrix

$$h = \frac{Eta}{2(1 - v)} \{\beta \ 1 \ -\beta \ 1 \ -\beta \ -1 \ \beta \ -1\} \tag{6}$$

where $\beta = b/a$, a is the plate width, and b is the plate height.

The thermal load $\mathbf{q}$ is given by

$$\mathbf{q} = \mathbf{h}\alpha T \tag{7}$$

where α is the coefficient of thermal expansion and T is the average temperature of the element. The thermal stiffness $\mathbf{h}$ represents the element forces needed to suppress thermal expansion αT.

For any dynamic analysis the element mass matrix $\mathbf{m}$ is used. For a rectangular plate the mass matrix must include inertia terms in the out-of-plane node displacements. This mass matrix is given by

$$\mathbf{m} = \frac{\rho abt}{36} \begin{bmatrix} 4 & 0 & 0 & 2 & 0 & 0 & 1 & 0 & 0 & 2 & 0 & 0 \\ 0 & 4 & 0 & 0 & 2 & 0 & 0 & 1 & 0 & 0 & 2 & 0 \\ 0 & 0 & 4 & 0 & 0 & 2 & 0 & 0 & 1 & 0 & 0 & 2 \\ 2 & 0 & 0 & 4 & 0 & 0 & 2 & 0 & 0 & 1 & 0 & 0 \\ 0 & 2 & 0 & 0 & 4 & 0 & 0 & 2 & 0 & 0 & 1 & 0 \\ 0 & 0 & 2 & 0 & 0 & 4 & 0 & 0 & 2 & 0 & 0 & 1 \\ 1 & 0 & 0 & 2 & 0 & 0 & 4 & 0 & 0 & 2 & 0 & 0 \\ 0 & 1 & 0 & 0 & 2 & 0 & 0 & 4 & 0 & 0 & 2 & 0 \\ 0 & 0 & 1 & 0 & 0 & 2 & 0 & 0 & 4 & 0 & 0 & 2 \\ 2 & 0 & 0 & 1 & 0 & 0 & 2 & 0 & 0 & 4 & 0 & 0 \\ 0 & 2 & 0 & 0 & 1 & 0 & 0 & 2 & 0 & 0 & 4 & 0 \\ 0 & 0 & 2 & 0 & 0 & 1 & 0 & 0 & 2 & 0 & 0 & 4 \end{bmatrix} \tag{8}$$

where a and b are the dimensions of the rectangular panel, t is the panel thickness, and ρ is the material density. This mass matrix is in three dimensions, including the normal direction to the plane of the plate.

1.2.6   Step 6: Computation of Element Properties in Global Coordinates

The next step after the element properties in local coordinates are computed is to convert these properties into the global coordinate system. These properties are denoted here by symbols with bars and are obtained using the λ matrices as shown next:

$$\bar{\mathbf{k}} = \lambda^T \mathbf{k} \lambda \tag{9}$$

$$\bar{\mathbf{h}} = \lambda^T \mathbf{h} \tag{10}$$

$$\bar{\mathbf{Q}} = \lambda^T \mathbf{Q} \tag{11}$$

$$\bar{\mathbf{m}} = \lambda^T \mathbf{m} \lambda \tag{12}$$

(See Sec. 1.3 for a discussion of the λ matrices.)

1.2.7   Step 7: Solution for the Global Node Displacements $\bar{\mathbf{U}}$

Once the properties of individual elements in the global coordinate system are calculated, they are then assembled into the complete structure using the global coordinate system. This is accomplished by a simple addition of the properties in the directions (i, j) to form the static load displacement equation for the whole structure as

$$\bar{K}\bar{U} = \bar{P} + \bar{Q} \tag{13}$$

where $\bar{K}$ is the stiffness matrix for the whole structure assembled from individual element stiffnesses, k^i is a column matrix representing external forces in the direction of $\bar{U}$, $\bar{U}$ and $\bar{P}$ are the externally applied loads, and $\bar{Q}$ are the thermal loads all in global coordinates.

1.2.8 Step 8: Computation of the Node Displacements

$\bar{K}$ is a singular matrix and therefore cannot be inverted to obtain a solution for $\bar{U}$. Therefore in this step all rows and columns corresponding to the rigid frame of reference (see step 3), and any other rows and columns corresponding to additional constraints (zero displacements) are eliminated from $\bar{K}$, $\bar{U}$, $\bar{P}$, and $\bar{Q}$. These new matrices are designated with subscripts r as $\bar{K}_r$, $\bar{U}_r$, $\bar{P}_r$, and $\bar{Q}_r$ to form the fundamental equation

$$\bar{K}_r \bar{U}_r = \bar{P}_r + \bar{Q}_r \tag{14}$$

Hence

$$\bar{U}_r = \bar{K}_r^{-1}(\bar{P}_r + \bar{Q}_r) \tag{15}$$

In practice, however, in order to preserve the original numbering system the rows and columns in $\bar{K}$ are not eliminated but are simply replaced with zeros except for the diagonal terms, which are replaced with ones. The justification for this process is explained as follows. The fundamental equation in any finite element analysis can be expressed as

$$\begin{bmatrix} \bar{K}_{1,1} & \bar{K}_{1,2} \\ \bar{K}_{2,1} & \bar{K}_{2,2} \end{bmatrix} \begin{bmatrix} \bar{U}_1 \\ \bar{U}_2 \end{bmatrix} = \begin{bmatrix} \bar{P}_1 + \bar{Q}_1 \\ \bar{P}_2 + \bar{Q}_2 \end{bmatrix} \tag{16}$$

where $\bar{U}_1$ represents unconstrained node displacements and $\bar{U}_2$ represents zero displacements at the rigid frame of reference and any other additional rigid constraints such as zero displacements to represent a condition of symmetry so that only one-half of the structure needs to be analyzed. Hence,

$$\begin{bmatrix} \bar{K}_{1,1} & 0 \\ 0 & I \end{bmatrix} \begin{bmatrix} \bar{U}_1 \\ \bar{U}_2 \end{bmatrix} = \begin{bmatrix} \bar{P}_1 + \bar{Q}_1 \\ 0 \end{bmatrix} \tag{17}$$

and

$$\begin{bmatrix} \bar{U}_1 \\ \bar{U}_2 \end{bmatrix} = \begin{bmatrix} \bar{K}_{1,1}^{-1} & 0 \\ 0 & I \end{bmatrix} \begin{bmatrix} \bar{P}_1 + \bar{Q}_1 \\ 0 \end{bmatrix} \tag{18}$$

If there is need to determine the reactions at the rigid frame of reference, this can be obtained from

$$\bar{\mathbf{P}}_2 + \bar{\mathbf{Q}}_2 = \bar{\mathbf{K}}_{21}\bar{\mathbf{U}}_1 \tag{19}$$

1.2.9 Step 9: Computation of the Element Local Stresses

The stresses in individual elements are obtained from the element stress equations in terms of the local displacements $\mathbf{u}_r$.

1.3 Transformation of Coordinate Axes: λ Matrices

To determine the stiffness property of the complete structure, a common datum must be established for all unassembled structural elements so that all the displacements and their corresponding forces will be referred to a common coordinate system. The choice of such a datum system is arbitrary, and in practice it is best selected to correspond to the coordinate system used on engineering drawings from which coordinates of different points on the structure can easily be found. Because the stiffness matrix $\mathbf{k}$ and the thermal stiffness $\mathbf{h}$ as well as the mass matrix $\mathbf{m}$ for each element are initially calculated in local coordinates, suitably oriented to minimize the computing effort, it is necessary to introduce transformation matrices changing the frame of reference from local coordinate system to global coordinate system (also referred to as the datum coordinate system). The first step in deriving such a transformation is to obtain a matrix relationship between the element displacements $\mathbf{u}$ in the local system and the element displacement $\bar{\mathbf{u}}$ in the datum system. This relationship is expressed by the matrix equation

$$\mathbf{u} = \lambda\bar{\mathbf{u}} \tag{20}$$

where λ is a matrix of coefficients obtained by resolving global displacements in the directions of local coordinates. It will be shown later that the elements of λ are obtained from the direction cosines of angles between the local and global coordinate systems.

If virtual displacements $\delta\bar{\mathbf{u}}$ are introduced on an element, then from Eq. (20)

$$\delta\mathbf{u} = \lambda\delta\bar{\mathbf{u}} \tag{21}$$

Because the resulting virtual work (a scalar quantity) must be independent of the coordinate system, it follows that

$$\delta\bar{\mathbf{u}}^T\bar{\mathbf{S}} = \delta\mathbf{u}^T\mathbf{S} \tag{22}$$

where $\bar{\mathbf{S}}$ refers to the element forces in global coordinate system corresponding to the displacements $\bar{\mathbf{u}}$. Substituting Eq. (21) into (22) leads to

$$\delta\bar{\mathbf{u}}^T\left(\bar{\mathbf{S}} - \lambda^T\mathbf{S}\right) = \mathbf{0} \tag{23}$$

and because $\delta\bar{\mathbf{u}}$ are arbitrary, it follows that

$$\bar{\mathbf{S}} - \lambda^T\mathbf{S} = \mathbf{0} \tag{24}$$

Using now Eqs. (20) and (24), the following element force-displacement equation is obtained in global coordinate system:

$$\bar{\mathbf{S}} = \bar{\mathbf{k}}\bar{\mathbf{u}} + \bar{\mathbf{h}}\alpha T = \bar{\mathbf{k}}\bar{\mathbf{u}} + \bar{\mathbf{q}} \tag{25}$$

where the element stiffness, element thermal stiffness, and element thermal loads in global coordinates are obtained from

$$\bar{\mathbf{k}} = \boldsymbol{\lambda}^T \mathbf{k} \boldsymbol{\lambda} \tag{26}$$

$$\bar{\mathbf{h}} = \boldsymbol{\lambda}^T \mathbf{h} \tag{27}$$

$$\bar{\mathbf{q}} = \boldsymbol{\lambda}^T \mathbf{h}\alpha T = \boldsymbol{\lambda}^T \mathbf{q} \tag{28}$$

The formulation of the transformation matrix $\boldsymbol{\lambda}$ will be illustrated for a pin-jointed bar element oriented arbitrarily in space, as shown in Fig. 12. The displacements in local coordinates can be related to those in global coordinates by the equations

$$u_p = \ell_{pq}\bar{u}_{3p-2} + m_{pq}\bar{u}_{3p-1} + n_{3y-1} + n_{pq}\bar{u}_{3p}$$
$$u_q = \ell_{pq}\bar{u}_{3q-2} + m_{pq}\bar{u}_{3q-1} + n_{3y-1} + n_{pq}\bar{u}_{3q} \tag{29}$$

where ℓ_{pq}, m_{pq}, and n_{pq} represent direction cosines of angles between the line pq and Ox, Oy, and Oz directions, respectively. Equations (29) can be arranged in

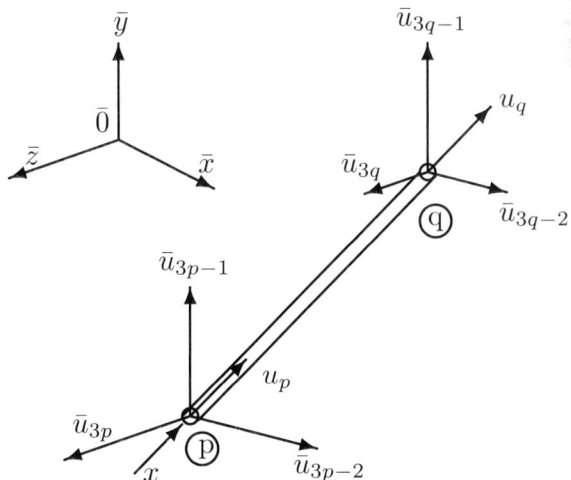

Fig. 12   Pin-jointed bar displacements in local and global coordinate systems.

matrix notation as

$$
\begin{bmatrix} u_p \\ u_q \end{bmatrix} = \begin{bmatrix} \ell_{pq} & m_{pq} & n_{pq} & 0 & 0 & 0 \\ 0 & 0 & 0 & \ell_{pq} & m_{pq} & n_{pq} \end{bmatrix} \begin{bmatrix} \bar{u}_{3p-2} \\ \bar{u}_{3p-1} \\ \bar{u}_{3p} \\ \bar{u}_{3q-2} \\ \bar{u}_{3q-1} \\ \bar{u}_{3q} \end{bmatrix}
\tag{30}
$$

Hence the transformation matrix λ is given by

$$
\lambda = \begin{bmatrix} \ell_{pq} & m_{pq} & n_{pq} & 0 & 0 & 0 \\ 0 & 0 & 0 & \ell_{pq} & m_{pq} & n_{pq} \end{bmatrix}
\tag{31}
$$

Substitution of Eq. (31) into Eqs. (26) and (27) leads finally to

$$
\bar{k} = \frac{AE}{\ell} \begin{bmatrix} k_0 & -k_0 \\ -k_0 & k_0 \end{bmatrix}
\tag{32}
$$

where

$$
k_0 = \begin{bmatrix} \ell_{pq}^2 & \ell_{pq}m_{pq} & \ell_{pq}m_{pq} \\ m_{pq}\ell_{pq} & m_{pq}^2 & m_{pq}n_{pq} \\ n_{pq}\ell_{pq} & n_{pq}m_{pq} & n_{pq}^2 \end{bmatrix}
\tag{33}
$$

and

$$
\bar{h} = AE\{\ell_{pq}m_{pq}n_{pq} - \ell_{pq} - m_{pq} - n_{pq}\}
\tag{34}
$$

Thus the matrix transformation given by Eq. (31) changes a (2×2) stiffness matrix k in a local coordinate system, measured along the length of the bar, into a (6×6) stiffness matrix $\bar{k}$ in the global coordinate system. Similarly, the transformation given by Eq. (34) changes a (2×1) matrix h into a (6×1) matrix $\bar{h}$. For the case of a pin-jointed bar discussed in this section, the λ matrix was a 2×6 matrix. In general, however, the λ matrix is $m \times n$, where m is the number of displacements (degrees of freedom) in local coordinates, while n is the number of displacements on the element in global coordinates. The displacements (degrees of freedom) are the actual displacements or rotations at the node points represented by vectors.

1.4 Basic Assumptions in Finite Element Analysis

The traditional method in finite element analysis is based on the assumed displacement field within each element. These fields satisfy equation of compatibility of strains, but in general they violate the equations of stress equilibrium within the element. Only the simple three-node triangular plate element from among the family of isoparametric elements does not violate the equations of stress equilibrium. Another method of determining the properties of finite elements is to use stress fields that satisfy equations of stress equilibrium, but this approach is limited only to some simple elements. For elements for which no stress fields are available, the displacement field can be augmented by an additional displacement field vanishing on the element boundaries and its magnitude determined from the principle of minimum potential energy in the element.

The fundamental equation underlying finite element analysis is the relationship between the displacements within the element (e.g., u, v, and w in the directions x, y, and z) and the node displacements. Mathematically this is expressed as

$$\mathbf{u} = \begin{bmatrix} u \\ v \\ w \end{bmatrix} = [\mathbf{a}] \begin{bmatrix} u_1 \\ u_2 \\ \vdots \\ u_n \end{bmatrix} \tag{35}$$

where $u_1, u_2, \ldots$, and u_n are the element node displacements. The matrix $\mathbf{a}$ is described as the *shape function* for the element. The shape functions for each element described in this text are given in Parts 2 and 3.

1.5 Upper and Lower Bounds in Finite Element Analysis

Finite element analysis per se does not provide any measure of accuracy of the results for any given representation of finite elements. Naturally as the number of elements is increased, the accuracy improves, converging to the true solution. This text describes a new method of assessing the accuracy of the finite element analysis by providing an upper and a lower bound for the solution. The closeness of the solution for the two bounds determines a measure of accuracy. The concept is very simple: the analysis is performed for two sets of elements. One set of elements is based on assumed displacement fields, which in general violate equations of stress equilibrium, whereas the other set is based on assumed stress fields for which the equations of stress equilibrium are satisfied. Only some simple elements based on displacement fields satisfy stress equilibrium (e.g., triangular plate). This concept is illustrated with a simple cantilever beam subjected to a transverse load at the tip in Figs. 13 and 14. The tip displacement and stress at the built-in end were computed for five different mesh idealizations ranging from one to 32 elements. The results presented are for two separate analyses using stress elements (T4) and displacement elements (T3), which provide the upper- and lower-bound solutions. As the number of elements is increased, the two solutions asymptotically approach the correct values of tip displacement and maximum stress in the beam. Also, the differences between the two solutions decrease as the number of elements is increased. The measure of accuracy of the results can be taken as

$$\text{Accuracy} = 1 - \frac{RS - RD}{RS} = \frac{RD}{RS} \tag{36}$$

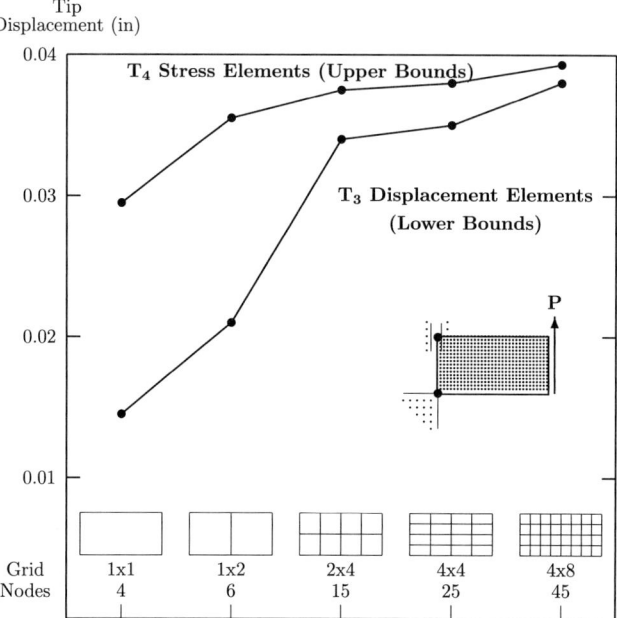

Fig. 13 Tip displacement in a cantilever under distributed load P for various grid sizes calculated with the stress and displacement elements: $P = 1000$ lb, $E = 10 \times 10^6$ lb/in.^{2}, $v = 0.3$, beam length = 24 in., beam height = 12 in., and thickness = 0.1 in.

where RD are the results for the assumed displacement field and RS are the results for the assumed stress field solution. The closeness of the ratio RD/RS to one is an indication of the accuracy for a given size of finite elements selected for the analysis. This is crucial information for analysts and engineering designers to decide whether the selected finite element mesh provides acceptable accuracy for a safe design. For some elements the stress field might not be available. For these cases the traditional displacement field is augmented by a displacement field vanishing on the boundaries, or subelements are used.

The diagrammatic representation of the upper- and lower-bounds analysis is shown in Fig. 15. Also, the concept of the lower- and upper-bound analysis cannot be used for simple elements such as pin-jointed bars, shear panels, and beam elements.

1.6   Equations of Stress Equilibrium for Finite Elements

It will be demonstrated that some of the elements in current use whose properties are based on the displacement distributions violate the stress equilibrium conditions within the element. This results in inaccuracies in the finite element results unless the grid size is sufficiently small.

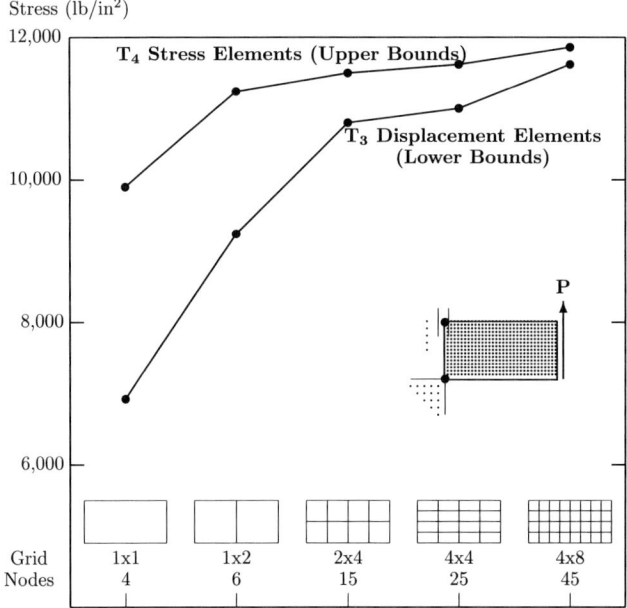

Fig. 14 Stress at the lower left corner in a cantilever under distributed load P for various grid sizes calculated with the stress and displacement elements: $P = 1000$ lb, $E = 10 \times 10^6$ lb/in.^{2}, $\nu = 0.3$, beam length = 24 in., beam height = 12 in., and thickness = 0.1 in.

The differential equations of stress equilibrium for two-dimensional elements are given by

$$\frac{\partial \sigma_{xx}}{\partial x} + \frac{\partial \sigma_{xy}}{\partial y} = 0 \tag{37}$$

$$\frac{\partial \sigma_{yy}}{\partial y} + \frac{\partial \sigma_{xy}}{\partial x} = 0 \tag{38}$$

These equations can be expressed in matrix notation as

$$\begin{bmatrix} \dfrac{\partial}{\partial x} & 0 & \dfrac{\partial}{\partial y} \\[2ex] 0 & \dfrac{\partial}{\partial y} & \dfrac{\partial}{\partial x} \end{bmatrix} \begin{bmatrix} \sigma_{xx} \\ \sigma_{yy} \\ \sigma_{xy} \end{bmatrix} = \begin{bmatrix} 0 \\ 0 \end{bmatrix} \tag{39}$$

To convert this equation into the node displacement relation, the stress-strain (Hooke's law) and the element strain-displacements equations are used, that is,

$$\begin{bmatrix} \sigma_{xx} \\ \sigma_{yy} \\ \sigma_{xy} \end{bmatrix} = \frac{E}{(1 - \nu^2)} \begin{bmatrix} 1 & \nu & 0 \\ \nu & 1 & 0 \\ 0 & 0 & (1-\nu)/2 \end{bmatrix} \begin{bmatrix} e_{xx} \\ e_{yy} \\ e_{xy} \end{bmatrix} \tag{40}$$

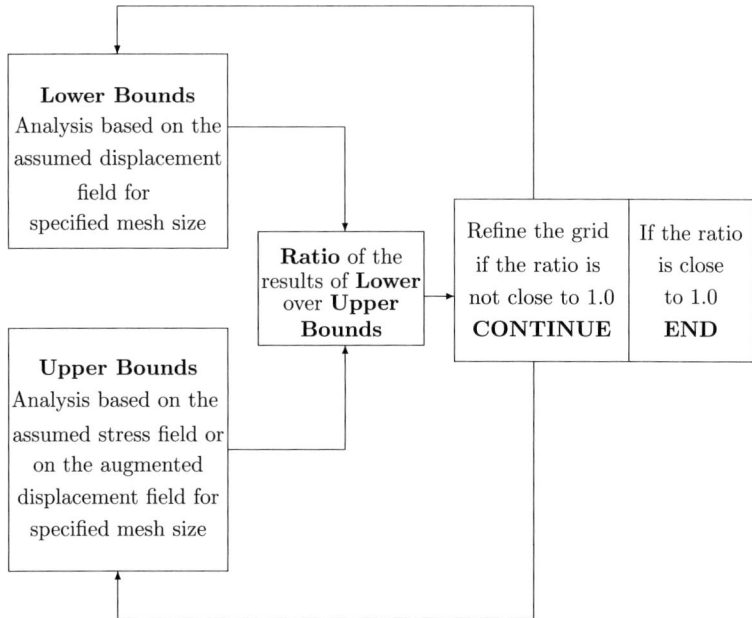

Fig. 15   Diagrammatic representation of the upper- and lower-bounds. (The ratio must be specified by the analyst, e.g., 0.98.)

and

$$\begin{bmatrix} e_{xx} \\ e_{yy} \\ e_{xy} \end{bmatrix} = \begin{bmatrix} \mathbf{b}_{xx} \\ \mathbf{b}_{yy} \\ \mathbf{b}_{xy} \end{bmatrix} \mathbf{u} \tag{41}$$

where $\mathbf{b}_{xx}$, $\mathbf{b}_{yy}$, and $\mathbf{b}_{xy}$ represent strains caused by unit displacements at the element nodes. Substitution of Eqs. (40) and (41) into the stress equilibrium equations [Eq. (39)] and canceling out the factor $E/(1 - v^2)$ lead to

$$\begin{bmatrix} \dfrac{\partial}{\partial x} & v\dfrac{\partial}{\partial x} & \dfrac{(1-v)}{2}\dfrac{\partial}{\partial y} \\ v\dfrac{\partial}{\partial y} & \dfrac{\partial}{\partial y} & \dfrac{(1-v)}{2}\dfrac{\partial}{\partial x} \end{bmatrix} \begin{bmatrix} \mathbf{b}_{xx} \\ \mathbf{b}_{yy} \\ \mathbf{b}_{xy} \end{bmatrix} \mathbf{u} = \begin{bmatrix} 0 \\ 0 \end{bmatrix} \tag{42}$$

Theoretically, the finite elements used in structural analysis should be based on displacement distributions that satisfy the preceding equations of equilibrium; however, this is not the case with most of the elements used today. This fact was recognized by John H. Argyris, one of the early pioneers of finite elements, that although the equations of stress equilibrium are violated, the effects are not pronounced as long as the element sizes are kept reasonably small. For triangular plate elements $\mathbf{b}_{xx}$, $\mathbf{b}_{yy}$, and $\mathbf{b}_{xy}$ contain only constant terms, which means that they satisfy Eq. (23); however, this is not the case for a rectangular plate element T3.

1.6.1 Rectangular Plate Element T3

For this element

$$\mathbf{b}_3 = \begin{bmatrix} \mathbf{b}_{xx} \\ \mathbf{b}_{yy} \\ \mathbf{b}_{xy} \end{bmatrix}$$

$$= \frac{1}{ab} \begin{bmatrix} -(b-y) & 0 & (b-y) & 0 & y & 0 & -y & 0 \\ 0 & -(a-x) & 0 & -x & 0 & x & 0 & (a-x) \\ -(a-x) & -(b-y) & -x & (b-y) & x & y & (a-x) & -y \end{bmatrix}$$

(43)

Substitution of Eq. (42) into the equations of stress equilibrium Eq. (42) leads to

$$\frac{1}{ab} \begin{bmatrix} 0 & \dfrac{(1+v)}{2} & 0 & \dfrac{(1+v)}{2} & 0 & \dfrac{(1+v)}{2} & 0 & \dfrac{(1+v)}{2} \\ \dfrac{(1+v)}{2} & 0 & \dfrac{(1+v)}{2} & 0 & \dfrac{(1+v)}{2} & 0 & \dfrac{(1+v)}{2} & 0 \end{bmatrix}$$

$$\times \begin{bmatrix} u_1 \\ \vdots \\ u_8 \end{bmatrix} = \begin{bmatrix} 0 \\ 0 \end{bmatrix}$$

which simplifies to

$$\frac{(1+v)}{2ab} \begin{bmatrix} 0 & 1 & 0 & -1 & 0 & 1 & 0 & -1 \\ 1 & 0 & -1 & 0 & 1 & 0 & -1 & 0 \end{bmatrix} \begin{bmatrix} u_1 \\ \vdots \\ u_8 \end{bmatrix} = \begin{bmatrix} 0 \\ 0 \end{bmatrix}$$

(44)

where $u_1 \ldots u_8$ are the node displacements on a rectangular plate (Fig. 16). The preceding equation is only satisfied if

$$u_2 - u_4 + u_6 - u_8 = 0$$

(45)

and

$$u_1 - u_3 + u_5 - u_7 = 0$$

(46)

which is true for rigid-body displacements (i.e., zero stresses) when $u_1 = u_3 = u_5 = u_7 =$ rigid-body displacement in the x direction, and $u_2 = u_4 = u_6 = u_8 =$ rigid-body displacement in the y direction, or when the element is subjected to uniform stretching when $u_7 = u_1$, $u_5 = u_3$, $u_8 = u_2$, and $u_6 = u_4$. In general, however, the stress equilibrium equations will not be satisfied, and this introduces inaccuracies in the solution.

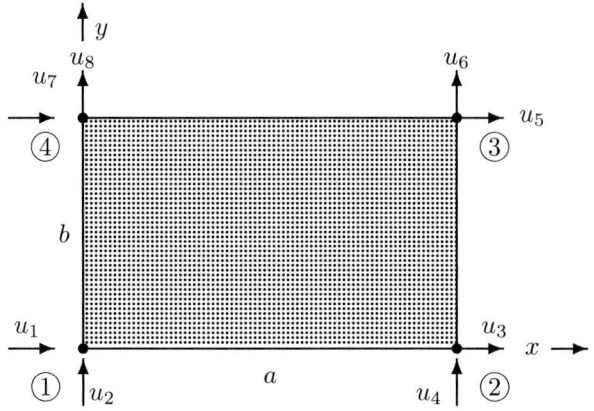

Fig. 16   Node displacements on rectangular plate element T3.

1.6.2   Rectangular Plate Element T4

For this element, $\mathbf{b}_4$ is given by

$$\mathbf{b}_4 = \frac{1}{2ab} \begin{bmatrix} -2(b-y) & v(a-2x) & 2(b-y) & -v(a-2x) & 2y \\ v(b-y) & -2(a-x) & -v(b-y) & -2x & v(b-2y) \\ -a & -b & -a & b & a \end{bmatrix}$$

$$\left. \begin{matrix} v(a-2x) & 2by & -v(a-2x) \\ 2x & -v(b-2y) & 2(a-x) \\ b & a & -b \end{matrix} \right] \tag{47}$$

Substituting now Eq. (47) into the equations of stress equilibrium, Eq. (42) leads to the following result:

$$\frac{1}{2ab} \begin{bmatrix} 0 & 0 & 0 & 0 & 0 & 0 & 0 & 0 \\ 0 & 0 & 0 & 0 & 0 & 0 & 0 & 0 \\ 0 & 0 & 0 & 0 & 0 & 0 & 0 & 0 \end{bmatrix} \begin{bmatrix} u_1 \\ \vdots \\ u_8 \end{bmatrix} = \begin{bmatrix} 0 \\ 0 \end{bmatrix} \tag{48}$$

which, of course, satisfies identically the equations of stress equilibrium.

1.6.3   Six-Node Isoparametric Triangular Plate Element

The six-node isoparametric element (Fig. 17) is introduced here only to show that it violates the equations of stress equilibrium within the element. This element is not used for the upper- and lower-bound analysis described in this text.

The assumed displacements for this isoparametric element are given by

$$u = c_1 + c_2 x + c_3 y + c_4 x^2 + c_5 xy + c_6 y^2 \tag{49}$$

$$v = c_7 + c_8 x + c_9 y + c_{10} x^2 + c_{11} xy + c_{12} y^2 \tag{50}$$

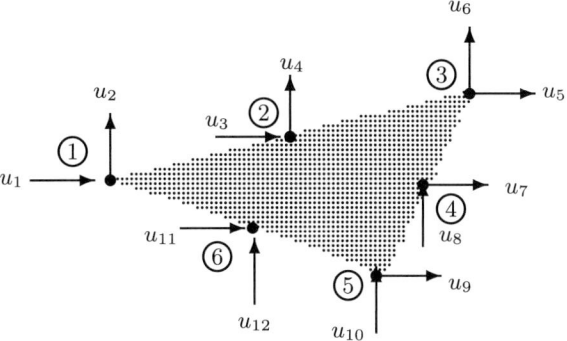

Fig. 17   Node displacements for a six-node isoparametric triangular plate element.

where $c_1, c_2 \cdots c_{12}$ are constants. The strains in the element are calculated from

$$
\begin{bmatrix} e_{xx} \\ e_{yy} \\ e_{xy} \end{bmatrix} = \begin{bmatrix} \partial u/\partial x \\ \partial v/\partial y \\ \partial u/\partial y + \partial v/\partial x \end{bmatrix}
$$

$$
= \begin{bmatrix} 0 & 1 & 0 & 2x & y & 0 & 0 & 0 & 0 & 0 & 0 & 0 \\ 0 & 0 & 0 & 0 & 0 & 0 & 0 & 0 & 1 & 0 & x & 2y \\ 0 & 0 & 1 & 0 & x & 2y & 0 & 1 & 0 & 2x & y & 0 \end{bmatrix} \begin{bmatrix} c_1 \\ \vdots \\ c_{12} \end{bmatrix} \tag{51}
$$

Using Eqs. (50) and (51), the node displacements $u_1 \cdots u_{12}$ can be calculated from

$$
\mathbf{u} = \begin{bmatrix} u_1 \\ u_2 \\ u_3 \\ u_4 \\ u_5 \\ u_6 \\ u_7 \\ u_8 \\ u_9 \\ u_{10} \\ u_{11} \\ u_{12} \end{bmatrix} = \begin{bmatrix} 1 & x_1 & y_1 & x_1^2 & x_1 y_1 & y_1^2 & 0 & 0 & 0 & 0 & 0 & 0 \\ 0 & 0 & 0 & 0 & 0 & 0 & 1 & x_1 & y_1 & x_1^2 & x_1 y_1 & y_1^2 \\ 1 & x_2 & y_2 & x_2^2 & x_2 y_2 & y_2^2 & 0 & 0 & 0 & 0 & 0 & 0 \\ 0 & 0 & 0 & 0 & 0 & 0 & 1 & x_2 & y_2 & x_2^2 & x_2 y_2 & y_2^2 \\ 1 & x_3 & y_3 & x_3^2 & x_3 y_3 & y_3^2 & 0 & 0 & 0 & 0 & 0 & 0 \\ 0 & 0 & 0 & 0 & 0 & 0 & 1 & x_3 & y_3 & x_3^2 & x_3 y_3 & y_3^2 \\ 1 & x_4 & y_4 & x_4^2 & x_4 y_4 & y_4^2 & 0 & 0 & 0 & 0 & 0 & 0 \\ 0 & 0 & 0 & 0 & 0 & 1 & 1 & x_4 & y_4 & x_4^2 & x_4 y_4 & y_4^2 \\ 1 & x_5 & y_5 & x_5^2 & x_5 y_5 & y_5^2 & 0 & 0 & 0 & 0 & 0 & 0 \\ 0 & 0 & 0 & 0 & 0 & 1 & 1 & x_5 & y_5 & x_5^2 & x_5 y_5 & y_5^2 \\ 1 & x_6 & y_6 & x_6^2 & x_6 y_6 & y_6^2 & 0 & 0 & 0 & 0 & 0 & 0 \\ 0 & 0 & 0 & 0 & 0 & 0 & 1 & x_6 & y_6 & x_6^2 & x_6 y_6 & y_6^2 \end{bmatrix} \begin{bmatrix} c_1 \\ c_2 \\ c_3 \\ c_4 \\ c_5 \\ c_6 \\ c_7 \\ c_8 \\ c_9 \\ c_{10} \\ c_{11} \\ c_{12} \end{bmatrix} \tag{52}
$$

Symbolically, the preceding equation can be represented as

$$
\mathbf{u} = \mathbf{Cc} \tag{53}
$$

where $\mathbf{C}$ is the 12×12 matrix of coefficients and

$$\mathbf{c} = \{c_1 \ c_2 \ \cdots \ u_{12}\} \tag{54}$$

Hence

$$\mathbf{c} = \mathbf{C}^{-1}\mathbf{u} \tag{55}$$

Using Eqs. (41), (51), and (55), it follows that

$$\begin{bmatrix} b_{xx} \\ b_{yy} \\ b_{xy} \end{bmatrix} \mathbf{u} = \begin{bmatrix} 0 & 1 & 0 & 2x & y & 0 & 0 & 0 & 0 & 0 & 0 & 0 \\ 0 & 0 & 0 & 0 & 0 & 0 & 0 & 0 & 1 & 0 & x & 2y \\ 0 & 0 & 1 & 0 & x & 2y & 0 & 1 & 0 & 2x & y & 0 \end{bmatrix} \times \mathbf{C}^{-1}\mathbf{u} \tag{56}$$

which when substituted into the equation of stress equilibrium (42) leads to

$$\begin{bmatrix} \dfrac{\partial}{\partial x} & v\dfrac{\partial}{\partial x} & \dfrac{(1-v)}{2}\dfrac{\partial}{\partial y} \\[3mm] v\dfrac{\partial}{\partial y} & \dfrac{\partial}{\partial y} & \dfrac{(1-v)}{2}\dfrac{\partial}{\partial x} \end{bmatrix} \begin{bmatrix} 0 & 1 & 0 & 2x & y & 0 & 0 & 0 & 0 & 0 & 0 & 0 \\ 0 & 0 & 0 & 0 & 0 & 0 & 0 & 0 & 1 & 0 & x & 2y \\ 0 & 0 & 1 & 0 & x & 2y & 0 & 1 & 0 & 2x & y & 0 \end{bmatrix}$$
$$\times \mathbf{C}^{-1}\mathbf{u} = \begin{bmatrix} 0 \\ 0 \end{bmatrix} \tag{57}$$

$$\begin{bmatrix} 0 & 0 & 0 & 2 & 0 & (1-v) & 0 & 0 & 0 & 0 & \dfrac{(1+v)}{2} & 0 \\[3mm] 0 & 0 & 0 & 0 & \dfrac{(1+v)}{2} & 0 & 0 & 0 & 0 & (1-v) & 0 & 2 \end{bmatrix}$$
$$\times \mathbf{C}^{-1}\mathbf{u} \neq \begin{bmatrix} 0 \\ 0 \end{bmatrix} \tag{58}$$

clearly demonstrating that the six-node isoparametric element violates the stress equilibrium equations.

Table 1 Finite element designation for the lower- and upper-bound analysis used in this text

Lower-bound elements (displacement elements)	Type	Upper-bound elements (stress, corrective, special elements)
T1	Triangular plate	T2 (corrective)
T3	Rectangular plate	T4 (stress)
T5	Quadrilateral plate	T6 (stress)
T7	Prismatic tetrahedron	T8 (corrective)
T9	Pentahedron	T10 (stress)
T11	Hexahedron	T12 (special)

1.7   Finite Elements Used for Lower- and Upper-Bound Analysis

The different types of finite elements used in this text are summarized in Table 1. Displacement elements use assumed displacement fields in terms of element displacements at specified nodes. This allows for the determination of strains and stresses within the element as well as element properties such as stiffness, thermal stiffness, and mass. Such elements are described as *displacement elements*, and they are used for the lower-bound analysis (T1, T3, T5, T7, T9, T11). The other category of elements used in this text are *stress elements* used for the upper-bound analysis (T4, T6, T10). For the stress elements, a stress field is assumed from which the corresponding displacement field is derived and element properties calculated. Unfortunately, such solutions are possible only for a few elements. A possible general method is to use additional displacement fields within the element but vanishing on the boundaries (described also as bubble elements). These additional fields have multiplying coefficients that are determined from the principle of minimum potential energy, and typically only one additional displacement field is used for u, v, and w displacements. This concept was proposed first by T. H. H. Pian in 1964. Examples of this approach used in this text are the triangular plate element T2 and the solid tetrahedron element T8. Element T12 in this text is a special hexahedron element made up from two prismatic tetrahedrons.

2
Static and Dynamic Properties of Finite Elements

2.1 Shape Functions

Shape functions that prescribe the displacement distributions within the element are used to derive the fundamental properties of finite elements. Such elements can therefore be described as *displacement elements*. For some elements, it is possible to derive element properties from an assumed stress distribution. An example of this is a rectangular plate element. Such elements can therefore be described as stress elements. For convenience, the elements with corrective displacement fields will also be treated as stress elements. For three-dimensional displacement elements, the shape functions are derived from the equation

$$\begin{bmatrix} u \\ v \\ w \end{bmatrix} = \begin{bmatrix} \mathbf{a}_{3D}(x, y, z) \end{bmatrix} \begin{bmatrix} u_1 \\ u_2 \\ \vdots \\ u_n \end{bmatrix} = \mathbf{a}_{3D}\mathbf{u} \tag{1}$$

where u, v, and w are displacements within the element in the x, y, and z directions and $\mathbf{u}$ is the column matrix of element displacements. Here $\mathbf{a}_{3D}$ is a matrix representing the shape function whose coefficients are assumed functions of x, y, and z and $u_1 \cdots u_n$ are displacements at the element nodes, where n is the number of element node displacements. For two-dimensional elements, only functions of x and y are used to define u and v displacements, whereas for one-dimensional elements only functions of x are used to define the u displacement.

Likewise, it can be demonstrated that for two- and one-dimensional elements the corresponding shape functions are given by

$$\begin{bmatrix} u \\ v \end{bmatrix} = \begin{bmatrix} \mathbf{a}_{2D}(x, y) \end{bmatrix} \begin{bmatrix} u_1 \\ u_2 \\ \vdots \\ u_n \end{bmatrix} = \mathbf{a}_{2D}\mathbf{u} \tag{2}$$

$$\begin{bmatrix} u \end{bmatrix} = \begin{bmatrix} \mathbf{a}_{1D}(x) \end{bmatrix} \begin{bmatrix} u_1 \\ u_2 \end{bmatrix} \mathbf{a}_{1D}\mathbf{u} \tag{3}$$

where the number of rows in $\mathbf{u}$ depends on the number of displacements in any given element.

2.2 Mass Matrices

The mass matrix $\mathbf{m}$ for a three-dimensional elements is determined from the integral

$$\mathbf{m} = \rho \int_{v} \mathbf{a}_{3D}^{T}\mathbf{a}_{3D}\,\mathrm{d}V \tag{4}$$

where ρ is the density of the element material, $\mathbf{a}_{3D}$ is the shape function and the integration is performed over the volume of the element. Likewise, for two-dimensional elements the mass matrix is determined from

$$\mathbf{m} = \rho \iint \mathbf{a}_{2D}^{T}\mathbf{a}_{2D}\,\mathrm{d}x\,\mathrm{d}y \tag{5}$$

and for one-dimensional elements from

$$\mathbf{m} = \rho \int \mathbf{a}_{1D}^{T}\mathbf{a}_{1D}\,\mathrm{d}x \tag{6}$$

2.3 Stiffness Matrices

The stiffness matrices for structural elements are derived from the matrix $\mathbf{b}$ in the relationship

$$\mathbf{e} = \mathbf{b}\mathbf{u} \tag{7}$$

where $\mathbf{e}$ is the matrix of strains in the element and $\mathbf{u}$ are the element displacements previously defined. The stiffness matrix for three-dimensional elements is calculated from the volume integral

$$\mathbf{k}_{3D} = \int_{v} \mathbf{b}_{3D}^{T}\mathbf{E}_{3}\mathbf{b}_{3D}\,\mathrm{d}V \tag{8}$$

where

$$\mathbf{E}_{3} = \frac{E}{(1+v)(1-2v)}\begin{bmatrix} 1-v & v & v & 0 & 0 & 0 \\ v & 1-v & v & 0 & 0 & 0 \\ v & v & 1-v & 0 & 0 & 0 \\ 0 & 0 & 0 & 1-2v & 0 & 0 \\ 0 & 0 & 0 & 0 & 1-2v & 0 \\ 0 & 0 & 0 & 0 & 0 & 1-2v \end{bmatrix} \tag{9}$$

E is the Young's modulus, v is the Poisson's ratio of the material; and the matrix $\mathbf{b}_{3D}$ is obtained from the strain-node displacement relationship of the form

$$\mathbf{e} = \mathbf{b}_{3D}\mathbf{u} \tag{10}$$

Also from Eq. (1),

$$\mathbf{c} = \mathbf{C}_{3D}^{-1}\mathbf{u} \tag{11}$$

The strains $\mathbf{e}$ in a three-dimensional element are computed from

$$\mathbf{e} = \begin{bmatrix} e_{xx} \\ e_{yy} \\ e_{zz} \\ e_{xy} \\ e_{yz} \\ e_{zx} \end{bmatrix} = \begin{bmatrix} \partial u/\partial x \\ \partial v/\partial y \\ \partial w/\partial z \\ \partial v/\partial x + \partial u/\partial y \\ \partial w/\partial y + \partial v/\partial z \\ \partial u/\partial z + \partial w/\partial x \end{bmatrix} = \mathbf{A}_{3D}\mathbf{c} = \mathbf{A}_{3D}\mathbf{C}^{-1}\mathbf{u} = \mathbf{b}_{3D}\mathbf{u} \tag{12}$$

where

$$\mathbf{b}_{3D} = \mathbf{A}_{3D}\mathbf{C}_{3D}^{-1} \tag{13}$$

in which $\mathbf{A}_{3D}$ is obtained from the partial derivatives of $\mathbf{F}$ with respect to x, y, and z.

The stiffness matrix for two-dimensional elements is calculated from the area integral

$$\mathbf{k}_{2D} = t \int_v \mathbf{b}_{2D}^T \mathbf{E}_2 \mathbf{b}_{2D} \, dx \, dy \tag{14}$$

where

$$\mathbf{E}_2 = \frac{E}{(1 - v^2)} \begin{bmatrix} 1 & v & 0 \\ v & 1 & 0 \\ 0 & 0 & (1 - v)/2 \end{bmatrix} \tag{15}$$

and t is the element thickness, E is the Young's modulus, v is the Poisson's ratio of the material, and the matrix $\mathbf{b}_{2D}$ is obtained from the strain-node displacement relationship shown next:

$$\mathbf{e} = \begin{bmatrix} e_{xx} \\ e_{yy} \\ e_{xy} \end{bmatrix} = \begin{bmatrix} \partial u/\partial x \\ \partial v/\partial y \\ \partial v/\partial x + \partial u/\partial y \end{bmatrix} = \mathbf{A}_{2D}\mathbf{c} = \mathbf{A}_{2D}\mathbf{C}_{2D}^{-1}\mathbf{u} = \mathbf{b}_{2D}\mathbf{u} \tag{16}$$

where

$$\mathbf{b}_{2D} = \mathbf{A}_{2D}\mathbf{C}_{2D}^{-1} \tag{17}$$

The stiffness matrix for the one-dimensional elements of length ℓ is calculated from the line integral

$$\mathbf{k}_{1D} = \int_0^\ell \mathbf{b}_{1D}^T E \mathbf{b}_{1D} \, dx \tag{18}$$

The strain-node displacement relationship is obtained from

$$e = \frac{\partial u}{\partial x} = A_{1D}c = AC_{1D}^{-1}u = b_{1D}u \tag{19}$$

where

$$b_{1D} = A_{1D}C_{1D}^{-1} \tag{20}$$

2.4 Thermal Stiffness Matrices

The element thermal stiffness for three-dimensional elements is computed from

$$h_{3D} = \int_v b_{3D}^T \frac{E}{(1-2v)} \begin{bmatrix} -1 \\ -1 \\ -1 \\ 0 \\ 0 \\ 0 \end{bmatrix} dV = \frac{EVb_{3D}^T}{(1-2v)} \begin{bmatrix} -1 \\ -1 \\ -1 \\ 0 \\ 0 \\ 0 \end{bmatrix} \tag{21}$$

where V is the volume of the element.
    For two-dimensional elements

$$h_{2D} = \int_V b_{2D}^T \frac{EV}{(1-v)} \begin{bmatrix} -1 \\ -1 \\ 0 \\ 0 \end{bmatrix} \tag{22}$$

For one-dimensional elements

$$h_{1D} = \int_0^1 b_{1D}^T EA\ell \, d\xi = EA \begin{bmatrix} -1 \\ 1 \end{bmatrix} \tag{23}$$

where A is the cross-sectional area of the element.

2.5 Thermal Load Matrices

The thermal loads for the elements are obtained from

$$q_{3D} = \alpha T h_{3D} \tag{24}$$
$$q_{2D} = \alpha T h_{2D} \tag{25}$$
$$q_{1D} = \alpha T h_{1D} \tag{26}$$

where α is the coefficient of thermal expansion and T is the element temperature, assumed to be uniform throughout the single element.

2.6 Static Equations

The static equation for a structure assembled from individual finite elements is represented by the equation

$$\mathbf{KU} = (\mathbf{P} + \mathbf{Q}) \tag{27}$$

where $\mathbf{K}$ is the assembled structure stiffness, $\mathbf{U}$ is the column matrix of the node displacements, $\mathbf{P}$ is the column matrix of applied loads at the nodes, and $\mathbf{Q}$ is the column matrix of the assembled thermal loads in the same directions as the loads $\mathbf{P}$. By eliminating the rigid-body degrees of freedom and any other additional rigid constraints, Eq. (27) becomes

$$\mathbf{K}_r \mathbf{U}_r = \mathbf{P}_r + \mathbf{Q}_r \tag{28}$$

Hence,

$$\mathbf{U}_r = \mathbf{K}_r^{-1}(\mathbf{P}_r + \mathbf{Q}_r) \tag{29}$$

2.7 Dynamic Equations

Equations of motion of a structure represented by finite elements are expressed as the matrix equation

$$\mathbf{M\ddot{U}} + \mathbf{C\dot{U}} + \mathbf{KU} = \mathbf{P}(t) + \mathbf{Q}(t) \tag{30}$$

where $\mathbf{M}$ is the assembled mass matrix for the whole structure (just like the assembled stiffness matrix $\mathbf{K}$), $\mathbf{C}$ represents the damping matrix, $\mathbf{\dot{U}}$ represents displacement velocities, and $\mathbf{\ddot{U}}$ represents accelerations in the directions of $\mathbf{U}$. Both $\mathbf{P}$ and $\mathbf{Q}$ are functions of time.

2.8 Natural Frequencies

Because the free oscillations are harmonic, the displacements $\mathbf{U}$ can be written as

$$\mathbf{U} = \mathbf{q}e^{-i\omega t} \tag{31}$$

where $\mathbf{q}$ is a column matrix of the amplitudes of the displacements $\mathbf{U}$, ω is the circular frequency of oscillations, and t is the time. Using Eq. (31) in Eq. (32) and then canceling the common factor $e^{i\omega t}$, it follows that

$$(-\omega^2 \mathbf{M} + \mathbf{K})\mathbf{q} = \mathbf{0} \tag{32}$$

which can be regarded as the equation of motion for the undamped freely oscillating system. Equation (33) has a nonzero solution for $\mathbf{q}$ provided

$$|-\omega^2 \mathbf{M} + \mathbf{K}| = 0 \tag{33}$$

The preceding equation is called the characteristic equation from which the natural frequencies of free oscillations ω can be calculated.

2.9 Dynamic Equations for Constrained Structures

If the rigid-body translations and rotations of the whole structure are eliminated, then the dynamic equation of motion and the characteristic equation for natural frequencies become

$$\mathbf{M}_r\ddot{\mathbf{U}}_r + \mathbf{C}_r\dot{\mathbf{U}}_r + \mathbf{K}_r\mathbf{U}_r = \mathbf{P}_r(t) + \mathbf{Q}_r(t) \tag{34}$$

and

$$|-\omega_r^2\mathbf{M}_r + \mathbf{K}_r| = 0 \tag{35}$$

Additional constraints beyond those used to suppress the rigid-body degrees of freedom can also be imposed.

2.10 Properties of Stress Elements and Enhanced Accuracy Elements

In general, the displacement elements violate the equations of stress equilibrium; one exception for which the stress equilibrium is satisfied is the triangular plate element. The determination of stress elements from an assumed stress distribution generally is difficult, and a more convenient approach is to superimpose onto the displacement element field an additional displacement field vanishing on the boundaries of the element and multiplied by a coefficient the magnitude of which is determined from the principle of minimum total potential energy in the element, for example, the T2 triangular plate element and T8 tetrahedron element.

Detailed discussion of the properties of the displacement elements, stress elements, and the enhanced accuracy elements can be found in Parts 2 and 3.

Part 2
Flat Plates

3
Triangular Plate T1: Assumed Displacement Distribution*

The node displacements for a triangular plate element $u_1 \cdots u_6$ and the node numbering are shown in Fig. 1.1. The element is subjected to element forces $S_1 \cdots S_6$ and thermal loads $q_1 \cdots q_6$ in the same directions as the corresponding element displacements $u_1 \cdots u_9$.

The displacement field in the x and y directions will be assumed as

$$u = c_1 + c_2 x + c_3 y \tag{1.1}$$

$$v = c_4 + c_5 x + c_6 y \tag{1.2}$$

where u and v are displacements in the x and y directions within the triangle and $c_1 \cdots c_6$ are constant coefficients. The preceding equations can be rewritten as

$$\begin{bmatrix} u \\ v \end{bmatrix} = \begin{bmatrix} 1 & x & y & 0 & 0 & 0 \\ 0 & 0 & 0 & 1 & x & y \end{bmatrix} \begin{bmatrix} c_1 \\ \vdots \\ c_6 \end{bmatrix} \tag{1.3}$$

Next, using Eq. (1.3), the element node displacements $\mathbf{u}$ can be expressed as

$$\mathbf{u} = \begin{bmatrix} u_1 \\ u_2 \\ u_3 \\ u_4 \\ u_5 \\ u_6 \end{bmatrix} = \begin{bmatrix} 1 & x_1 & y_1 & 0 & 0 & 0 \\ 0 & 0 & 0 & 1 & x_1 & y_1 \\ 1 & x_2 & y_2 & 0 & 0 & 0 \\ 0 & 0 & 0 & 1 & x_2 & y_2 \\ 1 & x_3 & y_3 & 0 & 0 & 0 \\ 0 & 0 & 0 & 1 & x_3 & y_3 \end{bmatrix} \begin{bmatrix} c_1 \\ c_2 \\ c_3 \\ c_4 \\ c_5 \\ c_6 \end{bmatrix} \tag{1.4}$$

Symbolically, the preceding equation can be expressed as

$$\mathbf{u} = \mathbf{C}_1 \mathbf{c}_1 \tag{1.5}$$

where $\mathbf{C}_1$ is the 6×6 matrix in Eq. (1.4), and therefore

$$\mathbf{c}_1 = \mathbf{C}_1^{-1} \mathbf{u} \tag{1.6}$$

---

*Finite elements used in this study are designated with a notation $\mathbf{Ti}$, where $\mathbf{i}$ is a number 1,2,3, etc. used as subscripts with matrix symbols denoting different element properties. Thus, $\mathbf{k}_3$ is used to denote matrix of stiffness coefficients for elements of type 3, which is a rectangular plate with the assumed displacements within its boundaries.

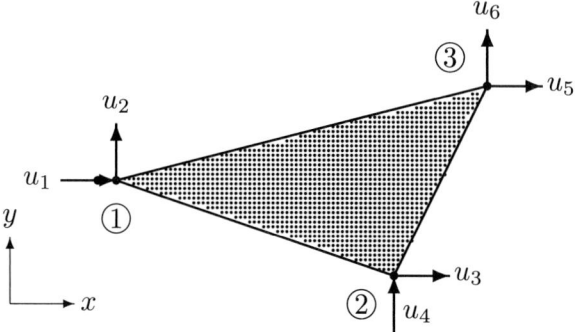

Fig. 1.1   Node displacements for triangular plate elements.

where

$$
\mathbf{C}_1^{-1} = \frac{1}{2A_{123}}
\begin{bmatrix}
(x_2y_3 - x_3y_2) & 0 & (-x_1y_3 + x_3y_1) \\
y_{23} & 0 & y_{13} \\
y_{23} & 0 & x_{13} \\
0 & (x_2y_3 - x_3y_2) & 0 \\
0 & y_{23} & 0 \\
0 & x_{32} & 0
\end{bmatrix}
$$

$$
\begin{bmatrix}
0 & (x_1y_2 - y_1x_2) & 0 \\
0 & y_{12} & 0 \\
0 & -x_{21} & 0 \\
(x_3y_1 - x_1y_3) & 0 & (x_1y_2 - y_1x_2) \\
-y_{13} & 0 & y_{12} \\
x_{13} & 0 & x_{12}
\end{bmatrix} \tag{1.7}
$$

and A_{123} is the area of the triangular plate given by

$$
A_{123} = \frac{1}{2}[(x_3 - x_1)(y_1 + y_3) + (x_2 - x_3)(+y_2 + y_3) + (x_2 - x_1)(y_1 + y_2)]/2
$$

$$
= (x_{12}y_{23} - x_{23}y_{12})/2 \tag{1.8}
$$

Substituting Eq. (1.6) into (1.3) leads to

$$
\begin{bmatrix} u \\ v \end{bmatrix} =
\begin{bmatrix} 1 & x & y & 0 & 0 & 0 \\ 0 & 0 & 0 & 1 & x & y \end{bmatrix}
\mathbf{C}_1^{-1}\mathbf{u} = \mathbf{a}_1\mathbf{u} \tag{1.9}
$$

where

$$
\mathbf{a}_1 =
\begin{bmatrix} 1 & x & y & 0 & 0 & 0 \\ 0 & 0 & 0 & 1 & x & y \end{bmatrix}
\mathbf{C}_1^{-1}
$$

$$
= \frac{1}{2A_{123}}
\begin{bmatrix}
y_{23} & 0 & y_{31} & 0 & y_{12} & 0 \\
0 & -x_{23} & 0 & -x_{31} & 0 & -x_{12} \\
-x_{23} & y_{23} & -x_{31} & y_{31} & -x_{12} & y_{12}
\end{bmatrix} \tag{1.10}
$$

represents the shape function derived from the assumed displacement field.

The strains $\mathbf{e}$ in the element are computed from

$$\mathbf{e} = \begin{bmatrix} e_{xx} \\ e_{yy} \\ e_{xy} \end{bmatrix} = \begin{bmatrix} \partial u/\partial x \\ \partial v/\partial y \\ \partial v/\partial x + \partial u/\partial y \end{bmatrix} = \begin{bmatrix} 0 & 1 & 0 & 0 & 0 & 0 \\ 0 & 0 & 0 & 0 & 0 & 1 \\ 0 & 0 & 1 & 0 & 1 & 0 \end{bmatrix} \begin{bmatrix} c_1 \\ \vdots \\ c_6 \end{bmatrix} \tag{1.11}$$

or symbolically as

$$\mathbf{e} = \mathbf{A}_1 \mathbf{c}_1 = \mathbf{A}_1 \mathbf{C}_1^{-1} \mathbf{u} = \mathbf{b}_1 \mathbf{u} \tag{1.12}$$

where $\mathbf{A}_1$ is the (3×6) matrix in Eq. (1.11) and

$$\mathbf{b}_1 = \mathbf{A}_1 \mathbf{C}_1^{-1} \tag{1.13}$$

Here the matrix $\mathbf{b}_1$ represents a matrix of strains caused by unit displacements at the element nodes. All three components of strains are constant throughout the element. By inverting $\mathbf{C}_1$, it can be shown that

$$\mathbf{b}_1 = \frac{1}{2A_{123}} \begin{bmatrix} y_{23} & 0 & y_{31} & 0 & y_{12} & 0 \\ 0 & -x_{23} & 0 & -x_{31} & 0 & -x_{12} \\ -x_{23} & y_{23} & -x_{31} & y_{31} & -x_{12} & y_{12} \end{bmatrix} \tag{1.14}$$

where A_{123} is the area of the triangle and

$$x_{ij} = x_i - x_j \tag{1.15}$$

$$y_{ij} = y_i - y_j \tag{1.16}$$

Equation (1.11) indicates that the assumption of linearly varying displacements within the triangular plate element leads to constant strains, and hence, by Hooke's Law, it also leads to constant stresses. This stress field satisfies the equations of strain compatibility as well as the equations of stress equilibrium.

The stress-strain equations are given by

$$\boldsymbol{\sigma} = \begin{bmatrix} \sigma_{xx} \\ \sigma_{yy} \\ \sigma_{xy} \end{bmatrix} = \frac{E}{(1 - v^2)} \begin{bmatrix} 1 & v & 0 \\ v & 1 & 0 \\ 0 & 0 & (1 - v)/2 \end{bmatrix} \begin{bmatrix} e_{xx} \\ e_{yy} \\ e_{xy} \end{bmatrix} - \frac{\alpha TE}{(1 - v)} \begin{bmatrix} 1 \\ 1 \\ 0 \end{bmatrix} \tag{1.17}$$

which can be written symbolically as

$$\boldsymbol{\sigma} = \mathbf{E}_2 \mathbf{e} - \frac{\alpha TE}{(1 - v)} \begin{bmatrix} 1 \\ 1 \\ 0 \end{bmatrix} = \mathbf{E}_2 \mathbf{b}_1 \mathbf{u} - \frac{\alpha TE}{(1 - v)} \begin{bmatrix} 1 \\ 1 \\ 0 \end{bmatrix} \tag{1.18}$$

where the two-dimensional Young's modulus $\mathbf{E}_2$ is given by

$$\mathbf{E}_2 = \frac{E}{(1 - v^2)} \begin{bmatrix} 1 & v & 0 \\ v & 1 & 0 \\ 0 & 0 & (1 - v)/2 \end{bmatrix} \tag{1.19}$$

Strains $\mathbf{e}$ are derived from the displacements caused by the external loading $\mathbf{P}$ and thermal loading $\mathbf{Q}$. The temperature T of the element will be assumed to be constant.

Element Stiffness

The stiffness matrix for the triangular plate is obtained from

$$\mathbf{k}_1 = \int_V \mathbf{b}_1^T \mathbf{E}_2 \mathbf{b}_1 \, dV = \iint \mathbf{b}_1^T \mathbf{E}_2 \mathbf{b}_1 t \, dx \, dy = \mathbf{b}_1^T \mathbf{E}_2 \mathbf{b}_1 t A_{123} \qquad (1.20)$$

where t is the plate thickness and A_{123} is the area of the triangle calculated from

$$A_{123} = \frac{1}{2} \begin{vmatrix} 1 & x_1 & y_1 \\ 1 & x_2 & y_2 \\ 1 & x_3 & y_3 \end{vmatrix} = \frac{1}{2}(x_{23}y_{31} - x_{31}y_{23}) \qquad (1.21)$$

Substituting Eqs. (1.14), (1.19), and (1.21) into Eq. (1.20) leads to

$$\mathbf{k}_1 = \frac{Et}{8A_{123}(1 - v^2)}$$

$$\times \begin{bmatrix}
2y_{23}^2 + (1-v)x_{23}^2 & -2vy_{23}x_{23} - (1-v)x_{23}y_{23} & 2y_{23}y_{31} + (1-v)x_{23}x_{31} \\
-2vx_{23}y_{23} - (1-v)y_{23}x_{23} & 2x_{23}^2 + (1-v)y_{23}^2 & -2vx_{23}y_{31} - (1-v)y_{23}x_{31} \\
2y_{31}y_{23} + (1-v)x_{31}x_{23} & -2vy_{31}x_{23} - (1-v)x_{31}y_{23} & 2y_{31}^2 + (1-v)x_{31}^2 \\
-2vx_{31}y_{23} - (1-v)y_{31}x_{23} & 2x_{23} + (1-v)y_{31}y_{23} & -2vx_{31}y_{31} - (1-v)y_{31}x_{31} \\
2y_{12}y_{23} + (1-v)x_{12}x_{23} & -2vy_{12}x_{23} - (1-v)x_{12}y_{23} & 2y_{12}y_{31} + (1-v)x_{12}y_{31} \\
-2vx_{12}y_{23} - (1-v)y_{12}x_{23} & 2x_{12}x_{23} + (1-v)y_{12}y_{23} & -2vy_{12}y_{31} - (1-v)y_{12}x_{31}
\end{bmatrix}$$

$$\begin{bmatrix}
-2vy_{23}x_{31} - (1-v)x_{23}y_{31} & 2y_{23}y_{12} + (1-v)x_{23}x_{12} & -2vy_{23}x_{12} - (1-v)x_{23}y_{12} \\
2x_{23}x_{31} + (1-v)y_{23}y_{31} & -2vx_{23}y_{12} - (1-v) - y_{23}x_{12} & 2x_{23}x_{12} + (1-v)y_{23}y_{12} \\
-2vy_{31}x_{31} - (1-v)x_{31}y_{31} & 2y_{31}y_{12} + (1-v)x_{31}x_{12} & -2vy_{31}x_{12} - (1-v)x_{31}y_{12} \\
2x_{31}^2 + (1-v)y_{31}^2 & -2vx_{31}y_{12} - (1-v)y_{31}x_{12} & 2x_{31}x_{12} + (1-v)y_{31}y_{12} \\
-2vy_{12}x_{31} - (1-v)x_{12}y_{31} & 2y_{12}^2 + (1-v)x_{12}^2 & -2vy_{12}x_{12} - (1-v)x_{12}y_{12} \\
2x_{12}x_{31} + (1-v)y_{12}y_{31} & -2vx_{12}y_{12} - (1-v)y_{12}x_{12} & 2x_{12}^2 + (1-v)y_{12}^2
\end{bmatrix}$$

$$(1.22)$$

As an example, the stiffness matrix $\mathbf{k}_1$ is shown in Eq. (1.23) for $v = 0.3$ and $x_1 = 0$, $y_1 = 0$, $x_2 = 10 \, \text{in.}$, $y_2 = 0$, $x_3 = 0$, and $y_3 = 10 \, \text{in.}$, that is, for a rectangular triangle.

$$\mathbf{k}_1 = Et \times \begin{bmatrix}
0.7418 & 0.3571 & -0.5495 & -0.1923 & -0.1923 & -0.1648 \\
0.3571 & 0.7418 & -0.1648 & -0.1923 & -0.1923 & -0.5495 \\
-0.5495 & -0.1648 & 0.5495 & 0 & 0 & 0.1648 \\
-0.1923 & -0.1923 & 0 & 0.1923 & 0.1923 & 0 \\
-0.1923 & -0.1923 & 0 & 0.1923 & 0.1923 & 0 \\
-0.1648 & -0.5495 & 0.1648 & 0 & 0 & 0.5495
\end{bmatrix}$$

$$(1.23)$$

The element thermal stiffness $\mathbf{h}_1$ is computed from

$$\mathbf{h}_1 = \int_V \mathbf{b}_1^T \frac{E}{(1-v)} \begin{bmatrix} -1 \\ -1 \\ 0 \end{bmatrix} dV = \frac{Et}{2(1-v)} \begin{bmatrix} -y_{23} \\ x_{23} \\ -y_{31} \\ x_{31} \\ -y_{12} \\ x_{12} \end{bmatrix} \tag{1.24}$$

Element Thermal Load

The element thermal load $\mathbf{q}_1$ is obtained from

$$\mathbf{q}_1 = \alpha T \mathbf{h}_1 \tag{1.25}$$

where T is the element temperature assumed to be constant within the element and α is the coefficient of thermal expansion.

Element Mass

To calculate the complete mass matrix for a plate element, we need not only the u and v displacements but also the w displacements. The element cannot resist loads in the z directions, and it will be assumed here that the displacement distribution in the z direction is of the same form as that for u or v, and the node displacements will be denoted with bars over the displacement symbols as shown in Fig. 1.2. Hence

$$w = c_7 + c_8 x + c_9 y \tag{1.26}$$

which when combined with Eqs. (1.1) and (1.2) leads to

$$\begin{bmatrix} u \\ v \\ w \end{bmatrix} = \begin{bmatrix} 1 & x & y & 0 & 0 & 0 & 0 & 0 & 0 \\ 0 & 0 & 0 & 1 & x & y & 0 & 0 & 0 \\ 0 & 0 & 0 & 0 & 0 & 0 & 1 & x & y \end{bmatrix} \begin{bmatrix} \bar{c}_1 \\ \vdots \\ \bar{c}_9 \end{bmatrix} = \bar{\mathbf{A}}_1 \bar{\mathbf{c}}_1 \tag{1.27}$$

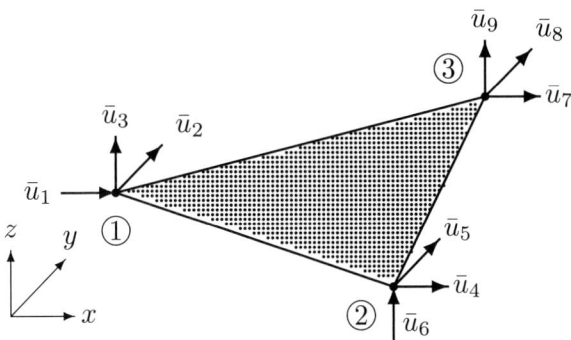

Fig. 1.2 Node displacements for use with the mass matrix for triangular plate elements.

where

$$\bar{\mathbf{c}}_1 = \bar{c}_1 \cdots \bar{c}_9 \tag{1.28}$$

and $\bar{\mathbf{A}}_1$ is the 3×9 matrix in Eq. (1.27). Hence, from Eq. (1.27)

$$\bar{\mathbf{u}} = \begin{bmatrix} u_1 \\ u_2 \\ u_3 \\ u_4 \\ u_5 \\ u_6 \\ u_7 \\ u_8 \\ u_9 \end{bmatrix} = \begin{bmatrix} 1 & x_1 & y_1 & 0 & 0 & 0 & 0 & 0 & 0 \\ 0 & 0 & 0 & 1 & x_1 & y_1 & 0 & 0 & 0 \\ 0 & 0 & 0 & 0 & 0 & 0 & 1 & x_1 & y_1 \\ 1 & x_2 & y_2 & 0 & 0 & 0 & 0 & 0 & 0 \\ 0 & 0 & 0 & 1 & x_2 & y_2 & 0 & 0 & 0 \\ 0 & 0 & 0 & 0 & 0 & 0 & 1 & x_2 & y_2 \\ 1 & x_3 & y_3 & 0 & 0 & 0 & 0 & 0 & 0 \\ 0 & 0 & 0 & 1 & x_3 & y_3 & 0 & 0 & 0 \\ 0 & 0 & 0 & 0 & 0 & 0 & 1 & x_3 & y_3 \end{bmatrix} \begin{bmatrix} \bar{c}_1 \\ \bar{c}_2 \\ \bar{c}_3 \\ \bar{c}_4 \\ \bar{c}_5 \\ \bar{c}_6 \\ \bar{c}_7 \\ \bar{c}_8 \\ \bar{c}_9 \end{bmatrix} = \bar{\mathbf{C}}_1 \bar{\mathbf{c}}_1 \tag{1.29}$$

where $\bar{\mathbf{C}}_1$ is the (9×9) matrix just given. It can be shown that its inverse is given by

$$\bar{\mathbf{C}}_1^{-1} = \frac{1}{2A_{123}} \begin{bmatrix} (y_2 x_3 - y_3 x_2) & 0 & 0 & (y_3 x_1 - y_1 x_3) & 0 \\ y_{32} & 0 & 0 & y_{13} & 0 \\ x_{23} & 0 & 0 & x_{31} & 0 \\ 0 & (y_2 x_3 - y_3 x_2) & 0 & 0 & (y_3 x_1 - y_1 x_3) \\ 0 & y_{32} & 0 & 0 & y_{13} \\ 0 & x_{23} & 0 & 0 & x_{31} \\ 0 & 0 & (y_2 x_3 - y_3 x_2) & 0 & 0 \\ 0 & 0 & y_{32} & 0 & 0 \\ 0 & 0 & x_{23} & 0 & 0 \end{bmatrix}$$

$$\begin{bmatrix} 0 & (y_1 x_2 - y_2 x_1) & 0 & 0 \\ 0 & y_{21} & 0 & 0 \\ 0 & x_{12} & 0 & 0 \\ 0 & 0 & (y_1 x_2 - y_2 x_1) & 0 \\ 0 & 0 & y_{21} & 0 \\ 0 & 0 & x_{12} & 0 \\ (y_3 x_1 - y_1 x_3) & 0 & 0 & (y_1 x_2 - y_2 x_1) \\ y_{13} & 0 & 0 & y_{21} \\ x_{31} & 0 & 0 & x_{12} \end{bmatrix} \tag{1.30}$$

From Eq. (1.29) it follows that

$$\bar{\mathbf{c}}_1 = \bar{\mathbf{C}}_1^{-1} \bar{\mathbf{u}} \tag{1.31}$$

and the shape function equation is obtained from Eq. (1.26) as

$$\begin{bmatrix} u \\ v \\ w \end{bmatrix} = \bar{\mathbf{A}}_1 \bar{\mathbf{C}}_1^{-1} \bar{\mathbf{u}} = \bar{\mathbf{a}}_1 \bar{\mathbf{u}} \tag{1.32}$$

where

$$\bar{\mathbf{a}}_1 = \bar{\mathbf{C}}_1 \bar{\mathbf{A}}_1^{-1} \tag{1.33}$$

$$\bar{\mathbf{u}}_1 = \{ u_1 \cdots u_9 \} \tag{1.34}$$

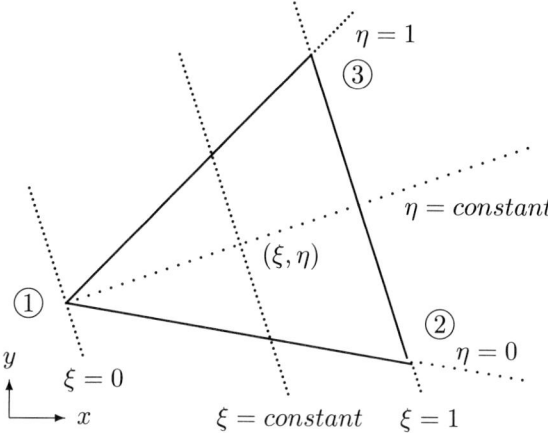

Fig. 1.3   Triangular coordinate system.

To determine the mass matrix, the resulting integration in local coordinates over the triangular area would be extremely unwieldy. Therefore it is preferable to introduce a special triangular coordinate system ξ and η shown in Fig. 1.3. These special coordinates are related to the rectangular coordinate system through the following equations:

$$x = x_1 - \xi(x_{12} + \eta x_{23}) \tag{1.35}$$

$$y = y_1 - \xi(y_{12} + \eta y_{23}) \tag{1.36}$$

Substituting the triangular coordinates ξ and η for x and y into Eq. (1.26) leads to a typical term for u as a function of u_1, u_4, and u_7:

$$u = (1 - \xi)u_1 + \xi(1 - \eta)u_4 + \xi\eta u_7 \tag{1.37}$$

With similar terms for the remaining displacements, it can be shown that

$$
\begin{bmatrix} u \\ v \\ w \end{bmatrix} =
\begin{bmatrix}
(1-\xi) & 0 & 0 & \xi\eta & 0 & 0 & \xi(1-\eta) & 0 & 0 \\
0 & (1-\xi) & 0 & 0 & \xi\eta & 0 & 0 & \xi(1-\eta) & 0 \\
0 & 0 & (1-\xi) & 0 & 0 & \xi\eta & 0 & 0 & \xi(1-\eta)
\end{bmatrix}
$$

$$\times \frac{1}{2A_{123}}\bar{\mathbf{u}} = \bar{\mathbf{a}}_1\bar{\mathbf{u}} \tag{1.38}$$

where $\bar{a}_1$ is the (3×9) matrix with the premultiplying factor $1/A_{123}$.

Denoting the element material density by ρ, the mass matrix is computed from

$$\bar{\mathbf{m}}_1 = \rho \iiint \bar{\mathbf{a}}_1^T \bar{\mathbf{a}}_1 \, dx \, dy \, dz = \rho t \int_0^1\!\!\int_0^1 \bar{\mathbf{a}}_1^T \bar{\mathbf{a}}_1 |J(x,y)| \, d\xi \, d\eta \tag{1.39}$$

where $|J(x, y)|$ is the determinant of the Jacobian J given by

$$|J(x, y)| = \frac{\partial x}{\partial \xi}\frac{\partial y}{\partial \eta} - \frac{\partial x}{\partial \eta}\frac{\partial y}{\partial \xi} = (x_{12} + \eta x_{23})y_{23}\xi - (y_{12} + \eta y_{23})x_{23}\xi$$

$$= (x_{12}y_{23} - x_{23}y_{12})\xi = 2A_{123}\xi \qquad (1.40)$$

Hence it can be shown that

$$\bar{\mathbf{m}}_1 = \frac{\rho A_{123}t}{12}\begin{bmatrix} 2 & 0 & 0 & 1 & 0 & 0 & 1 & 0 & 0 \\ 0 & 2 & 0 & 0 & 1 & 0 & 0 & 1 & 0 \\ 0 & 0 & 2 & 0 & 0 & 1 & 0 & 0 & 1 \\ 1 & 0 & 0 & 2 & 0 & 0 & 1 & 0 & 0 \\ 0 & 1 & 0 & 0 & 2 & 0 & 0 & 1 & 0 \\ 0 & 0 & 1 & 0 & 0 & 2 & 0 & 0 & 1 \\ 1 & 0 & 0 & 1 & 0 & 0 & 2 & 0 & 0 \\ 0 & 1 & 0 & 0 & 1 & 0 & 0 & 2 & 0 \\ 0 & 0 & 1 & 0 & 0 & 1 & 0 & 0 & 2 \end{bmatrix} \qquad (1.41)$$

4
Triangular Plate T2: Assumed Basic Displacement Distribution Plus Corrective Distribution Inside the Element Boundaries

The node numbering and element displacements $u_1 \cdots u_6$ for a triangular plate are shown in Fig. 2.1. At each node point there are two element forces and two displacements in the x and y directions, respectively. For each displacement there is a corresponding element force $S_1 \cdots S_6$.

The displacement field used for the T1 element (triangular plate) was represented by

$$u = c_1 + c_2 x + c_3 y \tag{2.1}$$

$$v = c_4 + c_5 x + c_6 y \tag{2.2}$$

where u and v are displacements at a point (x, y) in the x and y directions, respectively. To simplify the analysis, the origin for the rectangular coordinates will be taken at the node 1 with the y axis coinciding with the element edge from node 1 to node 3. The displacements will be evaluated in terms of the triangular coordinates shown also in Fig. 2.1. The x and y coordinates are functions of the ξ and η triangular coordinates as shown here:

$$x = \xi(1 - \eta)x_2 \tag{2.3}$$

and

$$y = \xi(y_2 - \eta y_{23}) \tag{2.4}$$

where

$$y_{23} = y_2 - y_3 \tag{2.5}$$

This leads then to the formulas for ξ and η as

$$\xi = \frac{x}{x_2(1 - \eta)} \tag{2.6}$$

and

$$\eta = \frac{1}{\xi y_{23}}(\xi y_2 - y) \tag{2.7}$$

43

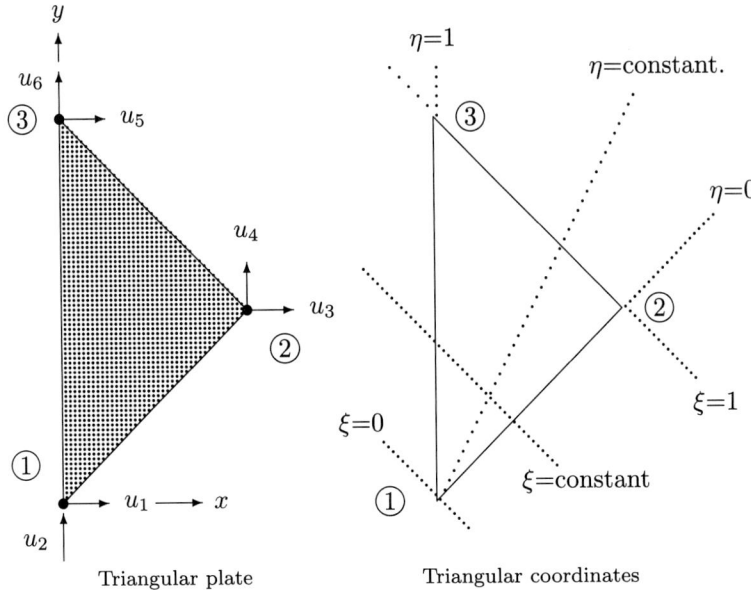

Triangular plate   Triangular coordinates

Fig. 2.1   Node displacements and triangular coordinates.

A simple function $f(\xi, \eta)$ that can be used as the corrective function for both u and v is given by

$$f(\xi, \eta) = \xi(1 - \xi)\eta(1 - \eta) \qquad (2.8)$$

This function vanishes on the boundaries of the triangular element. Thus the new displacement field for the triangle becomes

$$u = c_1 + c_2 x + c_3 y + c_7 f(\xi, \eta) \qquad (2.9)$$
$$v = c_4 + c_5 x + c_6 y + c_8 f(\xi, \eta) \qquad (2.10)$$

which can be represented as

$$\begin{bmatrix} u \\ v \end{bmatrix} = \begin{bmatrix} \mathbf{G}_a & \mathbf{G}_b \end{bmatrix} \mathbf{c} = \mathbf{Gc} \qquad (2.11)$$

where $\mathbf{G}$ is a rectangular matrix whose elements are the assumed displacement functions. The matrix $\mathbf{G}$ consists of two submatrices $\mathbf{G}_a$ and $\mathbf{G}_b$ given by

$$\mathbf{G}_a = \begin{bmatrix} 1 & x & y & 0 & 0 & 0 \\ 0 & 0 & 0 & 1 & x & y \end{bmatrix} \qquad (2.12)$$

and

$$\mathbf{G}_b = \begin{bmatrix} f(\xi, \eta) \\ f(\xi, \eta) \end{bmatrix} \qquad (2.13)$$

while the column matrix $\mathbf{c}$ is given by

$$\mathbf{c} = \{c_1 \; c_2 \; \cdots \; c_8\} = \{\mathbf{c}_a \; \mathbf{c}_b\} \tag{2.14}$$

where

$$\mathbf{c}_a = \{c_1 \cdots c_6\} \tag{2.15}$$

$$\mathbf{c}_b = \{c_7 \; c_8\} \tag{2.16}$$

Using Eqs. (2.9) and (2.10), the individual node displacements $\mathbf{u}$ can be expressed as

$$\mathbf{u} = \begin{bmatrix} u_1 \\ u_2 \\ u_3 \\ u_4 \\ u_5 \\ u_6 \end{bmatrix} = \begin{bmatrix} 1 & 0 & 0 & 0 & 0 & 0 & \vdots & 0 & 0 \\ 0 & 0 & 0 & 1 & 0 & 0 & \vdots & 0 & 0 \\ 1 & x_2 & y_2 & 0 & 0 & 0 & \vdots & 0 & 0 \\ 0 & 0 & 0 & 1 & x_2 & y_2 & \vdots & 0 & 0 \\ 1 & 0 & y_3 & 0 & 0 & 0 & \vdots & 0 & 0 \\ 0 & 0 & 0 & 1 & 0 & y_3 & \vdots & 0 & 0 \end{bmatrix} \begin{bmatrix} c_1 \\ c_2 \\ c_3 \\ c_4 \\ c_5 \\ c_6 \\ c_7 \\ c_8 \end{bmatrix} \tag{2.17}$$

Symbolically Eq. (2.17) can be expressed as

$$\mathbf{u} = \begin{bmatrix} \mathbf{C}_a & 0 \end{bmatrix} \begin{bmatrix} \mathbf{c}_a \\ \mathbf{c}_b \end{bmatrix} \tag{2.18}$$

where $\mathbf{C}_a$ is the 6×6 submatrix from Eq. (2.17). Hence it follows that

$$\mathbf{c}_a = \mathbf{C}_a^{-1} \mathbf{u} \tag{2.19}$$

where the inverse of $\mathbf{C}_a$ is given by

$$\mathbf{C}_a^{-1} = \begin{bmatrix} 1 & 0 & 0 & 0 & 0 & 0 \\ y_{23}/x_2 y_3 & 0 & -1/x_2 & 0 & -y_2/x_2 y_3 & 0 \\ -1/y_3 & 0 & 0 & 0 & 1/y_3 & 0 \\ 0 & 1 & 0 & 0 & 0 & 0 \\ 0 & y_{23}/x_2 y_3 & 0 & 1/x_2 & 0 & -y_2/x_2 y_3 \\ 0 & -1/y_3 & 0 & 0 & 0 & 1/y_3 \end{bmatrix} \tag{2.20}$$

where

$$x_{ij} = x_i - x_j \tag{2.21}$$

$$y_{ij} = y_i - y_j \tag{2.22}$$

To obtain the element strains, partial derivatives of the displacements must be obtained first. To find the partial derivatives, the chain rule of differentiation is used as given here:

$$\frac{\partial}{\partial x} = \frac{\partial}{\partial \xi} \frac{\partial \xi}{\partial x} + \frac{\partial}{\partial \eta} \frac{\partial \eta}{\partial x} \tag{2.23}$$

and

$$\frac{\partial}{\partial y} = \frac{\partial}{\partial \xi}\frac{\partial \xi}{\partial y} + \frac{\partial}{\partial \eta}\frac{\partial \eta}{\partial y} \tag{2.24}$$

From the relationship between the rectangular and the triangular coordinates in Eqs. (2.6) and (2.7), it follows that

$$\frac{\partial \xi}{\partial y} = 0 \tag{2.25}$$

and

$$\frac{\partial \eta}{\partial x} = 0 \tag{2.26}$$

Hence from Eq. (2.8),

$$\frac{\partial f}{\partial x} = \frac{\partial f}{\partial \xi}\frac{\partial \xi}{\partial x} = \frac{(1 - 2\xi)\eta}{x_2} \tag{2.27}$$

and

$$\frac{\partial f}{\partial y} = \frac{\partial f}{\partial \eta}\frac{\partial \eta}{\partial y} = \frac{-(1 - \xi)(1 - 2\eta)}{y_{23}} \tag{2.28}$$

The strains in the element can now be computed from

$$e_{xx} = \frac{\partial u}{\partial x} = c_2 + \frac{c_7(1 - 2\xi)\eta}{x_2} \tag{2.29}$$

$$e_{yy} = \frac{\partial v}{\partial y} = c_6 - \frac{c_8(1 - \xi)(1 - 2\eta)}{y_{23}} \tag{2.30}$$

$$e_{xy} = \frac{\partial u}{\partial y} + \frac{\partial v}{\partial x} = c_3 - \frac{c_7(1 - 2\eta)(1 - \xi)}{y_{23}} + \frac{c_8(1 - 2\xi)\eta}{x_2} \tag{2.31}$$

which leads to

$$e = \begin{bmatrix} e_{xx} \\ e_{yy} \\ e_{xy} \end{bmatrix} = Hc = \begin{bmatrix} H_a & H_b \end{bmatrix}\begin{bmatrix} c_a \\ c_b \end{bmatrix} \tag{2.32}$$

where

$$H = \begin{bmatrix} H_a & H_b \end{bmatrix} \tag{2.33}$$

$$H_a = \begin{bmatrix} 0 & 1 & 0 & 0 & 0 & 0 \\ 0 & 0 & 0 & 0 & 0 & 1 \\ 0 & 0 & 1 & 0 & 1 & 0 \end{bmatrix} \tag{2.34}$$

$$H_b = \begin{bmatrix} (1 - 2\xi)\eta/x_2 & 0 \\ 0 & -(1 - \xi)(1 - 2\eta)/y_{23} \\ -(1 - \xi)(1 - 2\eta)/y_{23} & (1 - 2\xi)\eta/x_2 \end{bmatrix} \tag{2.35}$$

Combining now Eq. (2.19) with the identity $\mathbf{c}_b = \mathbf{c}_b$ yields

$$\mathbf{c} = \begin{bmatrix} \mathbf{c}_a \\ \mathbf{c}_b \end{bmatrix} = \begin{bmatrix} \mathbf{C}_a^{-1} & \mathbf{0} \\ \mathbf{0} & \mathbf{I} \end{bmatrix} \begin{bmatrix} \mathbf{u} \\ \mathbf{c}_b \end{bmatrix} = \mathbf{W}\hat{\mathbf{u}} \tag{2.36}$$

where

$$\mathbf{W} = \begin{bmatrix} \mathbf{C}_a^{-1} & \mathbf{0} \\ \mathbf{0} & \mathbf{I} \end{bmatrix} = \begin{bmatrix} \mathbf{W}_{11} & \mathbf{W}_{12} \\ \mathbf{W}_{21} & \mathbf{W}_{22} \end{bmatrix} \tag{2.37}$$

$$\mathbf{W}_{11} = \mathbf{C}_a^{-1} \tag{2.38}$$

$$\mathbf{W}_{12} = \mathbf{0} \tag{2.39}$$

$$\mathbf{W}_{21} = \mathbf{0} \tag{2.40}$$

$$\mathbf{W}_{22} = \mathbf{I} \tag{2.41}$$

and

$$\hat{\mathbf{u}} = \begin{bmatrix} \mathbf{u} \\ \mathbf{c}_b \end{bmatrix} \tag{2.42}$$

The strain energy U_i in the element is given by

$$U_i = \frac{1}{2} \int_v \mathbf{e}^T \mathbf{E}_2 \mathbf{e}^T \, \mathrm{d}V = \frac{1}{2} \mathbf{c}^T \int_v \mathbf{H}^T \mathbf{E}_2 \mathbf{H} \, \mathrm{d}V \mathbf{c}$$

$$= \frac{1}{2} \hat{\mathbf{u}}^T \mathbf{W}^T \int_v \mathbf{H}^T \mathbf{E}_3 \mathbf{H} \, \mathrm{d}V \mathbf{W} \hat{\mathbf{u}} = \frac{1}{2} \hat{\mathbf{u}}^T \hat{\mathbf{k}} \hat{\mathbf{u}} \tag{2.43}$$

where

$$\hat{\mathbf{k}} = \mathbf{W}^T \int_v \mathbf{H}^T \mathbf{E}_2 \mathbf{H} \, \mathrm{d}V \mathbf{W} = \begin{bmatrix} \mathbf{k}_{aa} & \mathbf{k}_{ab} \\ \mathbf{k}_{ba} & \mathbf{k}_{bb} \end{bmatrix} \tag{2.44}$$

and

$$\mathbf{E}_2 = \frac{E}{(1 - \nu^2)} \begin{bmatrix} 1 & \nu & 0 \\ \nu & 1 & 0 \\ 0 & 0 & (1 - \nu)/2 \end{bmatrix} \tag{2.45}$$

The volume integral in Eq. (2.44) is calculated from

$$\int_v \mathbf{H}^T \mathbf{E}_2 \mathbf{H} \, \mathrm{d}V = t \int_{\eta=0}^1 \int_{\xi=0}^1 \mathbf{H}^T \mathbf{E}_2 \mathbf{H} |J(\xi, \eta)| \, \mathrm{d}\xi \, \mathrm{d}\eta \tag{2.46}$$

where t is the element thickness and $|J(\xi, \eta)|$ is the determinant of the Jacobian J given by

$$|J(x, y)| = \frac{\partial x}{\partial \xi} \frac{\partial y}{\partial \eta} - \frac{\partial x}{\partial \eta} \frac{\partial y}{\partial \xi} = (x_{12} + \eta x_{23})y_{23}\xi - (y_{12} + \eta y_{23})x_{23}\xi$$

$$= (x_{12}y_{23} - x_{23}y_{12})\xi = 2A_{123}\xi \tag{2.47}$$

and A_{123} is the area of the triangular plate.

The total potential energy U in the element can be written as

$$U = U_i - \mathbf{u}^T\mathbf{S}$$

$$= \frac{1}{2}\hat{\mathbf{u}}^T\hat{\mathbf{k}}\hat{\mathbf{u}} - \mathbf{u}^T\mathbf{S} = \frac{1}{2}\begin{bmatrix} \mathbf{u}^T & \mathbf{c}_b^T \end{bmatrix}\begin{bmatrix} \mathbf{k}_{aa} & \mathbf{k}_{ab} \\ \mathbf{k}_{ba} & \mathbf{k}_{bb} \end{bmatrix}\begin{bmatrix} \mathbf{u} \\ \mathbf{c}_b \end{bmatrix} - \begin{bmatrix} \mathbf{u}^T & \mathbf{c}_b^T \end{bmatrix}\begin{bmatrix} \mathbf{S} \\ 0 \end{bmatrix} \quad (2.48)$$

where $\mathbf{S}$ is a column matrix of the element forces corresponding with the displacements $\mathbf{u}$ and the term $\mathbf{u}^T\mathbf{S}$ is the potential of external forces. The condition of minimum potential energy requires that

$$\frac{\partial U}{\partial \hat{\mathbf{u}}} = 0 \quad (2.49)$$

leading to

$$\begin{bmatrix} \mathbf{k}_{aa} & \mathbf{k}_{ab} \\ \mathbf{k}_{ba} & \mathbf{k}_{bb} \end{bmatrix}\begin{bmatrix} \mathbf{u} \\ \mathbf{c}_b \end{bmatrix} - \begin{bmatrix} \mathbf{S} \\ 0 \end{bmatrix} = \begin{bmatrix} 0 \\ 0 \end{bmatrix} \quad (2.50)$$

The matrix $\mathbf{c}_b$ can be expressed in terms of $\mathbf{u}$ by solving the second row of the n equations in Eq. (2.50) so that

$$\mathbf{c}_b = -\mathbf{k}_{bb}^{-1}\mathbf{k}_{ba}\mathbf{u} \quad (2.51)$$

which when substituted into the first row of the m equations in Eq. (2.50) results in

$$(\mathbf{k}_{aa} - \mathbf{k}_{ab}\mathbf{k}_{bb}^{-1}\mathbf{k}_{ba})\mathbf{u} = \mathbf{S} \quad (2.52)$$

Hence, by definition, the element stiffness matrix $\mathbf{k}_2$ is given by

$$\mathbf{k}_2 = \mathbf{k}_{aa} - \mathbf{k}_{ab}\mathbf{k}_{bb}^{-1}\mathbf{k}_{ba} \quad (2.53)$$

The component stiffnesses $\mathbf{k}_{aa}, \mathbf{k}_{bb}, \mathbf{k}_{ab}$, and $\mathbf{k}_{ba}$ can be extracted from Eq. (2.44). Alternatively, these components can be obtained by matrix operations as shown next for $\mathbf{k}_{aa}$:

$$\mathbf{k}_{aa} = \begin{bmatrix} \mathbf{I}_{(6\times6)} & \mathbf{0}_{(6\times2)} \end{bmatrix}\begin{bmatrix} \mathbf{k}_{aa} & \mathbf{k}_{ab} \\ \mathbf{k}_{ba} & \mathbf{k}_{bb} \end{bmatrix}\begin{bmatrix} \mathbf{I}_{(6\times6)} \\ \mathbf{0}_{(6\times2)} \end{bmatrix} \quad (2.54)$$

where the subscripts with the identity matrices $\mathbf{I}$ and the zero matrices $\mathbf{0}$ denote the numbers of rows $\times$ the number of columns.

As an example, the stiffness matrix $\mathbf{k}_2$ is shown for $v = 0.3$ and $x_1 = 0, y_1 = 0$, $x_2 = 10\,\text{in.}, y_2 = 0, x_3 = 0$, and $y_3 = 10\,\text{in.}$, that is, for a rectangular triangle.

$$\mathbf{k}_2 = Et \times \begin{bmatrix} 0.675 & 0.325 & -0.50 & -0.175 & -0.175 & -0.15 \\ 0.325 & 0.675 & -0.15 & -0.175 & -0.175 & -0.50 \\ -0.500 & -0.15 & 0.50 & 0 & 0 & 0.15 \\ -0.175 & -0.175 & 0 & 0.175 & 0.175 & 0 \\ -0.175 & -0.175 & 0 & 0.175 & 0.175 & 0 \\ -0.15 & -0.50 & 0.15 & 0 & 0 & 0.50 \end{bmatrix} \quad (2.55)$$

The equation for the shape function, that is, a relationship between the displacements u and v within the element and the node displacements $\mathbf{u}$, can be obtained from Eq. (2.11) as

$$\begin{bmatrix} u \\ v \end{bmatrix} = \begin{bmatrix} 1 & x & y & 0 & 0 & 0 & \vdots & f(\xi,\eta) \\ 0 & 0 & 0 & 1 & x & y & \vdots & f(\xi,\eta) \end{bmatrix} \begin{bmatrix} \mathbf{c}_a \\ \mathbf{c}_b \end{bmatrix}$$

$$= \begin{bmatrix} 1 & x & y & 0 & 0 & 0 & \vdots & f(\xi,\eta) \\ 0 & 0 & 0 & 1 & x & y & \vdots & f(\xi,\eta) \end{bmatrix} \begin{bmatrix} \mathbf{C}_a^{-1} \\ -\mathbf{k}_{bb}^{-1}\mathbf{k}_{ba} \end{bmatrix} \mathbf{u} \qquad (2.56)$$

In the preceding equation $f(\xi,\eta)$ is a function of x and y.

The strains $\mathbf{e}$ are now determined from Eq. (2.32), which can be rewritten as

$$\mathbf{e} = \mathbf{H}_a\mathbf{c}_a + \mathbf{H}_b\mathbf{c}_b \qquad (2.57)$$

The equation for $\mathbf{c}_a$ from Eq. (2.36) can be rewritten as

$$\mathbf{c}_a = \mathbf{C}_a^{-1} \qquad (2.58)$$

and using $\mathbf{c}_b$ from Eq. (2.47) in Eq. (2.32), the strains $\mathbf{e}$ are computed from

$$\mathbf{e} = (\mathbf{H}_a\mathbf{C}_a^{-1} - \mathbf{H}_b\mathbf{k}_{bb}^{-1})\mathbf{u} \qquad (2.59)$$

The corresponding stresses are then obtained from

$$\sigma = \mathbf{E}_2\mathbf{e} \qquad (2.60)$$

Thermal Stiffness, Thermal Loads, and Mass Matrix

As an approximation, the thermal stiffness, thermal loads, and mass can be assumed to be the same as for the T1 element, that is,

$$\mathbf{h}_2 \approx \mathbf{h}_1 \qquad (2.61)$$

$$\mathbf{q}_2 \approx \mathbf{q}_1 \qquad (2.62)$$

$$\mathbf{m}_2 \approx \mathbf{m}_1 \qquad (2.63)$$

Corrective Functions for the Assumed Displacements

The corrective displacements $f(\xi,\eta) = \xi(1-\xi)\eta(1-\eta)$ [see Eq. (2.8)] vanish on the boundaries of the element creating a bubble over the triangular plate element. This type of element can therefore be described as a *bubble element*.

Other elements discussed in this text, which can also be described as the bubble elements, are the T8 tetrahedron and the T10 prismatic pentahedron.

5

Rectangular Plate T3: Assumed Displacement Distribution

The node displacements for rectangular plate elements and the node numbering are shown in Fig. 3.1. The displacements u and v within the element will be assumed as

$$u = c_1\xi + c_2\xi\eta + c_3\eta + c_4 \tag{3.1}$$

$$v = c_5\xi + c_6\xi\eta + c_7\eta + c_8 \tag{3.2}$$

where $\xi = x/a$, $\eta = y/b$, and arbitrary constants $c_1, \ldots, c_8$ are determined from the known node displacements in the x and y directions $(u_1, \ldots, u_8)$. The assumed displacement distribution is usually referred to as the bipolar distribution. The preceding equations can be rewritten as

$$
\begin{bmatrix} u \\ v \end{bmatrix} = \begin{bmatrix} \xi & \xi\eta & \eta & 1 & 0 & 0 & 0 & 0 \\ 0 & 0 & 0 & 0 & -\xi & \xi\eta & \eta & 1 \end{bmatrix} \begin{bmatrix} c_1 \\ \vdots \\ c_8 \end{bmatrix} \tag{3.3}
$$

Using Eqs. (3.1) and (3.2), the element node displacements $\mathbf{u}$ can be expressed as

$$
\mathbf{u} = \begin{bmatrix} u_1 \\ u_2 \\ u_3 \\ u_4 \\ u_5 \\ u_6 \\ u_7 \\ u_8 \end{bmatrix} = \begin{bmatrix} 0 & 0 & 0 & 1 & 0 & 0 & 0 & 0 \\ 0 & 0 & 0 & 0 & 0 & 0 & 0 & 1 \\ 1 & 0 & 0 & 1 & 0 & 0 & 0 & 0 \\ 0 & 0 & 0 & 0 & 1 & 0 & 0 & 1 \\ 1 & 1 & 1 & 1 & 0 & 0 & 0 & 0 \\ 0 & 0 & 0 & 0 & 1 & 1 & 1 & 1 \\ 0 & 0 & 1 & 1 & 0 & 0 & 0 & 0 \\ 0 & 0 & 0 & 0 & 0 & 0 & 1 & 1 \end{bmatrix} \begin{bmatrix} c_1 \\ c_2 \\ c_3 \\ c_4 \\ c_5 \\ c_6 \\ c_7 \\ c_8 \end{bmatrix} \tag{3.4}
$$

Symbolically the preceding equation can be expressed as

$$\mathbf{u} = \mathbf{C}_3\mathbf{c}_3 \tag{3.5}$$

where $\mathbf{C}_3$ is the (8×8) matrix in Eq. (3.4) and $\mathbf{c}_3 = \{c_1 \cdots c_8\}$. Hence

$$\mathbf{c}_3 = \mathbf{C}_3^{-1}\mathbf{u} \tag{3.6}$$

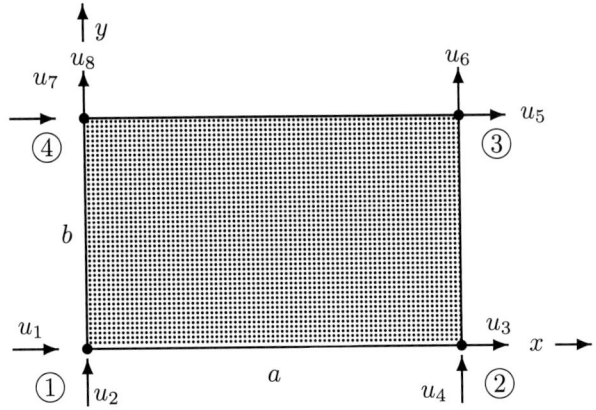

Fig. 3.1   Node displacements for rectangular plate elements.

Substituting Eq. (3.6) into Eq. (3.3) leads to the shape function equation

$$\begin{bmatrix} u \\ v \end{bmatrix} = \begin{bmatrix} (1-\xi)(1-\eta) & 0 & \xi(1-\eta) & 0 & \xi\eta & 0 \\ 0 & (1-\xi)(1-\eta) & 0 & \xi(1-\eta) & 0 & \xi\eta \end{bmatrix}$$

$$\begin{matrix} \eta(1-\xi) & 0 \\ 0 & \eta(1-\xi) \end{matrix} \bigg] \begin{bmatrix} u_1 \\ \vdots \\ u_8 \end{bmatrix}$$

$$= \mathbf{a}_3\mathbf{u} \tag{3.7}$$

where $\mathbf{a}_3$ represents the shape functions for this element.

The strains $\mathbf{e}$ in the element are then computed from Eqs. (3.1) and (3.2) as follows:

$$\mathbf{e} = \begin{bmatrix} e_{xx} \\ e_{yy} \\ e_{xy} \end{bmatrix} = \begin{bmatrix} \partial u/\partial x \\ \partial v/\partial y \\ \partial v/\partial x + \partial u/\partial y \end{bmatrix} = \begin{bmatrix} \dfrac{1}{a}\partial u/\partial \xi \\ \dfrac{1}{b}\partial v/\partial \eta \\ \dfrac{1}{a}\partial v/\partial \xi + \dfrac{1}{b}\partial u/\partial \eta \end{bmatrix} = \mathbf{A}_3\mathbf{c}_3 \tag{3.8}$$

where $\mathbf{A}_3$ is obtained by the indicated partial differentiation of $\mathbf{a}_3$. Hence symbolically $\mathbf{e}$ is obtained from

$$\mathbf{e} = \mathbf{A}_3\mathbf{c}_3 = \mathbf{A}_3\mathbf{C}_3^{-1}\mathbf{u} = \mathbf{b}_3\mathbf{u} \tag{3.9}$$

where

$$\mathbf{b}_3 = \mathbf{A}_3\mathbf{C}_3^{-1} \tag{3.10}$$

Here the matrix $\mathbf{b}_3$ represents a matrix of strains caused by unit displacements at the element nodes. By inverting $\mathbf{C}_3$, it can be shown that

$$
\mathbf{b}_3 = \frac{1}{ab}
\begin{bmatrix}
-b(1-\eta) & 0 & b(1-\eta) & 0 & b\eta \\
0 & -a(1-\xi) & 0 & -a\xi & 0 \\
-a(1-\xi) & -b(1-\eta) & -a\xi & b(1-\eta) & a\xi
\end{bmatrix}
$$

$$
\begin{bmatrix}
0 & -b\eta & 0 \\
a\xi & 0 & a(1-\xi) \\
b\eta & a(1-\xi) & -b\eta
\end{bmatrix}
\tag{3.11}
$$

The preceding equation indicates that the assumption of linear distribution of displacements within the rectangular plate element leads to linearly varying strains, and hence, by Hooks' Law, it also leads to linearly varying stresses. The stress field satisfies the equations of strain compatibility, but, it does not satisfy the equations of stress equilibrium within the element. The stress-strain equations are given by

$$
\boldsymbol{\sigma} = \begin{bmatrix} \sigma_{xx} \\ \sigma_{yy} \\ \sigma_{xy} \end{bmatrix} = \frac{E}{(1-v^2)} \begin{bmatrix} 1 & v & 0 \\ v & 1 & 0 \\ 0 & 0 & (1-v)/2 \end{bmatrix} \begin{bmatrix} e_{xx} \\ e_{yy} \\ e_{xy} \end{bmatrix} - \frac{\alpha TE}{(1-v)} \begin{bmatrix} 1 \\ 1 \\ 0 \end{bmatrix}
\tag{3.12}
$$

which can be written symbolically as

$$
\boldsymbol{\sigma} = \mathbf{E}_2 \mathbf{e} - \frac{\alpha TE}{(1-v)} \{1\ 1\ 0\}
\tag{3.13}
$$

where the two-dimensional Young's modulus is given by

$$
\mathbf{E}_2 = \frac{E}{(1-v^2)} \begin{bmatrix} 1 & v & 0 \\ v & 1 & 0 \\ 0 & 0 & (1-v)/2 \end{bmatrix}
\tag{3.14}
$$

Strains $\mathbf{e}$ are derived from the displacements caused by the external loading $\mathbf{P}$ and thermal loading $\mathbf{Q}$. The temperature T of the element can be constant or a function of the x and y coordinates.

Element Stiffness

The stiffness matrix for the rectangular plate element is obtained from

$$
\mathbf{k}_3 = \int_V \mathbf{b}_3^T \mathbf{E}_2 \mathbf{b}_3 \, dV = t \int_0^b \int_0^a \mathbf{b}_3^T \mathbf{E}_2 \mathbf{b}_3 \, dx \, dy = tab \int_0^1 \int_0^1 \mathbf{b}_3^T \mathbf{E}_2 \mathbf{b}_3 \, d\xi \, d\eta
\tag{3.15}
$$

where t is the panel thickness (assumed constant).

The stiffness matrix $\mathbf{k}_3$ for a rectangular plate element derived from Eq. (3.15) is given by

$$
\mathbf{k}_3 = \begin{bmatrix} \mathbf{k}_{11} & \mathbf{k}_{12} \\ \mathbf{k}_{21} & \mathbf{k}_{22} \end{bmatrix}
\tag{3.16}
$$

where

$$\mathbf{k}_{11} = \frac{Et}{12(1 - v^2)}$$

$$\times \begin{bmatrix} 4\beta + 2(1 - v)/\beta & 3(1 + v)/2 & -4\beta + (1 + v)/\beta & -3(1 - 3v)/2 \\ 3(1 + v)/2 & 4/\beta + 2(1 - v)\beta & 3(1 - 3v)/2 & 2/\beta - 2(1 - v)\beta \\ -4\beta + (1 + v)/\beta & 3(1 - 3v)/2 & 4\beta + 2(1 - v)/\beta & -3(1 + v)/2 \\ 3(1 - 3v)/2 & 2/\beta - 2(1 - v)\beta & -3(1 + v)/2 & 4\beta + 2(1 - v)\beta \end{bmatrix}$$

$$(3.17)$$

$$\mathbf{k}_{21} = \frac{Et}{12(1 - v^2)}$$

$$\times \begin{bmatrix} -2\beta - (1 - v)/\beta & -3(1 + v)/2 & 2\beta - 2(1 - v)/\beta & 3(1 - 3v)/2 \\ -3(1 + v)/2 & -2/\beta - (1 - v)\beta & -3(1 - 3v)/2 & -4/\beta + (1 - v)\beta \\ 2\beta - 2(1 - v)/\beta & -3(1 - 3v)/2 & -2\beta - (1 - v)/\beta & 3(1 + v)/2 \\ 3(1 - 3v)/2 & -4/\beta + (1 - v)\beta & 3(1 + v)/2 & -2/\beta - (1 - v)\beta \end{bmatrix}$$

$$(3.18)$$

$$\mathbf{k}_{12} = \mathbf{k}_{21}^{T} \qquad\qquad (3.19)$$

$$\mathbf{k}_{22} = \frac{Et}{12(1 - v^2)}$$

$$\times \begin{bmatrix} 4\beta + 2(1 - v)/\beta & 3(1 + v)/2 & -4\beta + (1 - v)/\beta & -3(1 - 3v)/2 \\ 3(1 + v)/2 & 4/\beta + 2(1 - v)\beta & 3(1 - 3v)/2 & 2/\beta - 2(1 - v)\beta \\ 4\beta + (1 - v)/\beta & 3(1 - 3v)/2 & 4\beta + 2(1 - v)/\beta & -3(1 + v)/2 \\ -3(1 - 3v)/2 & 2/\beta - 2(1 - v)\beta & -3(1 + v)/2 & 4/\beta + 2(1 - v)\beta \end{bmatrix}$$

$$(3.20)$$

For $\beta = 1$ and $v = 0.3$, the stiffness matrix $\mathbf{k}_3$ is given by

$$\mathbf{k}_3 = Et \begin{bmatrix} 0.4945 & 0.1786 & -0.3022 & -0.0137 & -0.2473 & -0.1786 & 0.0549 & 0.0137 \\ 0.1786 & 0.4945 & 0.0137 & 0.0549 & -0.1786 & -0.2473 & -0.0137 & -0.3022 \\ -0.3022 & 0.0137 & 0.4945 & -0.1786 & 0.0549 & -0.0137 & -0.2473 & 0.1786 \\ -0.0137 & 0.0549 & -0.1786 & 0.4945 & 0.0137 & -0.3022 & 0.1786 & -0.2473 \\ -0.2473 & -0.1786 & 0.0549 & 0.0137 & 0.4945 & 0.1786 & -0.3022 & -0.0137 \\ -0.1786 & -0.2473 & -0.0137 & -0.3022 & 0.1786 & 0.4945 & 0.0137 & 0.0549 \\ 0.0549 & -0.0137 & -0.2473 & 0.1786 & -0.3022 & 0.0137 & 0.4945 & -0.1786 \\ 0.0137 & -0.3022 & 0.1786 & -0.2473 & -0.0137 & 0.0549 & -0.1786 & 0.4945 \end{bmatrix}$$

$$(3.21)$$

Solving for the constants $c_1, \ldots, c_8$ from Eqs. (3.3) and (3.5), the element displacement field can be expressed in terms of the node displacements as

$$\begin{bmatrix} u \\ v \end{bmatrix} = \begin{bmatrix} (1 - \xi)(1 - \eta) & 0 & (1 - \xi)\eta & 0 & \xi\eta & 0 & \xi(1 - \eta) \\ 0 & (1 - \xi)(1 - \eta) & 0 & (1 - \xi) & \xi\eta & 0 & \xi(1 - \eta) \end{bmatrix}$$

$$\times \begin{bmatrix} u_1 \\ \vdots \\ u_8 \end{bmatrix} \qquad\qquad (3.22)$$

where the function of ξ and η in the preceding matrix are usually referred to as the shape functions. Now, using the Hooke's law for the plane stress problems, the stress-displacement equation becomes

$$
\begin{bmatrix} \sigma_{xx} \\ \sigma_{yy} \\ \sigma_{xy} \end{bmatrix} = \frac{E}{(1-v^2)ab} \left[\begin{array}{ccc} -b(1-\eta) & -va(1-\xi) & b(1-\eta) \\ -vb(1-\eta) & -a(1-\xi) & vb(1-\eta) \\ -(1-v)a(1-\xi)/2 & -(1-v)b(1-\eta)/2 & -(1-v)a\xi/2 \end{array} \right.
$$

$$
\begin{array}{ccc} -va\xi & b\eta & va\xi \\ -a\xi & vb\eta & a\xi \\ (1-v)b(1-\eta)/2 & (1-v)a\xi/2 & (1-v)b\eta/2 (1-v)a(1-\xi)/2 \end{array}
$$

$$
\left. \begin{array}{cc} -b\eta & va(1-\xi) \\ -vb\eta & a(1-\xi) \\ -(1-v)b\eta/2 & \end{array} \right] \begin{bmatrix} u_1 \\ \vdots \\ u_8 \end{bmatrix} \tag{3.23}
$$

Stress Equilibrium

Using Eq. (3.21), it follows that

$$
\frac{\partial \sigma_{xx}}{\partial x} = \frac{1}{a}\frac{\partial \sigma_{xx}}{\partial \xi} = \frac{E}{(1-v^2)ab} \begin{bmatrix} 0 & v & 0 & -v & 0 & v & 0 & -v \end{bmatrix} \mathbf{u} \tag{3.24}
$$

$$
\frac{\partial \sigma_{yy}}{\partial y} = \frac{1}{b}\frac{\partial \sigma_{yy}}{\partial \eta} = \frac{E}{(1-v^2)ab} \begin{bmatrix} v & 0 & -v & 0 & v & 0 & -v & 0 \end{bmatrix} \mathbf{u} \tag{3.25}
$$

$$
\frac{\partial \sigma_{xy}}{\partial x} = \frac{1}{a}\frac{\partial \sigma_{xy}}{\partial \xi}
$$

$$
= \frac{E}{(1-v^2)ab} \left[\frac{(1-v)}{2} \quad 0 \quad -\frac{(1-v)}{2} \quad 0 \quad \frac{(1-v)}{2} \quad 0 \quad \frac{(1-v)}{2} \quad 0 \right] \mathbf{u} \tag{3.26}
$$

$$
\frac{\partial \sigma_{xy}}{\partial y} = \frac{1}{b}\frac{\partial \sigma_{xy}}{\partial \eta}
$$

$$
= \frac{E}{(1-v^2)ab} \left[0 \quad \frac{(1-v)}{2} \quad 0 \quad -\frac{(1-v)}{2} \quad 0 \quad \frac{(1-v)}{2} \quad 0 \quad -\frac{(1-v)}{2} \right] \mathbf{u} \tag{3.27}
$$

where

$$
\mathbf{u} = \{u_1 \cdots u_8\} \tag{3.28}
$$

Hence the two equations of stress equilibrium within the element can be rewritten as

$$
\begin{bmatrix} \dfrac{\partial \sigma_{xx}}{\partial x} + \dfrac{\partial \sigma_{xy}}{\partial y} \\ \dfrac{\partial \sigma_{yy}}{\partial y} + \dfrac{\partial \sigma_{xy}}{\partial x} \end{bmatrix} = \frac{E}{2(1-v^2)ab} \begin{bmatrix} 0 & 1 & 0 & -1 & 0 & 1 & 0 & -1 \\ 1 & 0 & -1 & 0 & 1 & 0 & -1 & 0 \end{bmatrix} \mathbf{u} = \begin{bmatrix} 0 \\ 0 \end{bmatrix} \tag{3.29}
$$

The preceding equations are only satisfied if

$$u_2 - u_4 + u_6 - u_8 = 0 \qquad (3.30a)$$

and

$$u_1 - u_3 + u_5 - u_7 = 0 \qquad (3.30b)$$

which is true for either rigid-body displacements (i.e., zero stresses) when $u_1 = u_3 = u_5 = u_7 =$ rigid-body displacement in the x direction, and $u_2 = u_4 = u_6 = u_8 =$ rigid-body displacement in the y direction, or when the element is subjected to uniform stretching when $u_7 = u_1, u_5 = u_3, u_8 = u_2,$ and $u_6 = u_4$. In general, the stress equilibrium equations will not be satisfied, and this introduces inaccuracies in the solution.

Thermal Stiffness

The thermal stiffness for this element is computed from

$$\mathbf{h}_3 = \int_v \mathbf{b}_3^T \frac{E}{(1-v)} \begin{bmatrix} -1 \\ -1 \\ 0 \end{bmatrix} dV = \frac{abtE}{(1-v)} \int_0^1 \int_0^1 \mathbf{b}_3^T \begin{bmatrix} -1 \\ -1 \\ 0 \end{bmatrix} d\xi\, d\eta$$

$$= \frac{Eta}{2(1-v)} \{\beta \quad 1 \quad -\beta \quad 1 \quad -\beta \quad -1 \quad \beta \quad -1\} \qquad (3.31)$$

Thermal Load

The thermal load for the element is obtained from

$$\mathbf{q}_3 = \alpha T \mathbf{h}_3 \qquad (3.32)$$

where T is the temperature of the element (assumed to be constant) and α is the coefficients of thermal expansion.

The thermal loads $\mathbf{q}_3$ represent a self-equilibrating force system needed to supress thermal expansion caused by the temperature increase T.

Element Mass

For the mass matrix three displacements will be assumed at each node point as shown in Fig. 3.2 The additional displacement w needed for the mass matrix $\mathbf{m}_3$ can be assumed to be of the same form as u and v, that is,

$$w = c_9\xi + c_{10}\xi\eta + c_{11}\eta + c_{12} \qquad (3.33)$$

Combining the three equations (3.1), (3.2), and (3.33) leads to

$$\begin{bmatrix} u \\ v \\ w \end{bmatrix} = \begin{bmatrix} \xi & \xi\eta & \eta & 1 & 0 & 0 & 0 & 0 & 0 & 0 & 0 & 0 \\ 0 & 0 & 0 & 0 & \xi & \xi\eta & \eta & 1 & 0 & 0 & 0 & 0 \\ 0 & 0 & 0 & 0 & 0 & 0 & 0 & 0 & \xi & \xi\eta & \eta & 1 \end{bmatrix} \begin{bmatrix} c_1 \\ c_2 \\ \vdots \\ c_{12} \end{bmatrix} = \bar{\mathbf{C}}_3 \bar{\mathbf{c}}_3$$

$$(3.34)$$

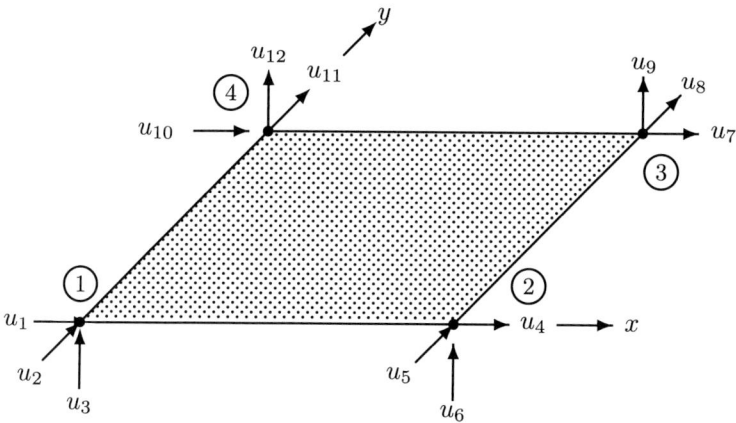

Fig. 3.2   Node displacements for the mass matrix m₃.

Hence

$$\bar{\mathbf{u}} = \begin{bmatrix} u_1 \\ u_2 \\ \vdots \\ u_{12} \end{bmatrix}$$

$$= \begin{bmatrix} 0 & 0 & 0 & 1 & 0 & 0 & 0 & 0 & 0 & 0 & 0 & 0 \\ 0 & 0 & 0 & 0 & 0 & 0 & 0 & 1 & 0 & 0 & 0 & 0 \\ 0 & 0 & 0 & 0 & 0 & 0 & 0 & 0 & 0 & 0 & 0 & 1 \\ 1 & 0 & 0 & 1 & 0 & 0 & 0 & 0 & 0 & 0 & 0 & 0 \\ 0 & 0 & 0 & 0 & 1 & 0 & 0 & 1 & 0 & 0 & 0 & 0 \\ 0 & 0 & 0 & 0 & 0 & 0 & 0 & 0 & 1 & 0 & 0 & 1 \\ 1 & 1 & 1 & 1 & 0 & 0 & 0 & 0 & 0 & 0 & 0 & 0 \\ 0 & 0 & 0 & 0 & 1 & 1 & 1 & 1 & 0 & 0 & 0 & 0 \\ 0 & 0 & 0 & 0 & 0 & 0 & 0 & 0 & 1 & 1 & 1 & 1 \\ 0 & 0 & 1 & 1 & 0 & 0 & 0 & 0 & 0 & 0 & 0 & 0 \\ 0 & 0 & 0 & 0 & 0 & 0 & 1 & 1 & 0 & 0 & 0 & 0 \\ 0 & 0 & 0 & 0 & 0 & 0 & 0 & 0 & 0 & 0 & 1 & 1 \end{bmatrix} \begin{bmatrix} c_1 \\ c_2 \\ \vdots \\ c_{12} \end{bmatrix} = \bar{\mathbf{A}}_3 \bar{\mathbf{c}}_3 \qquad (3.35)$$

where $\mathbf{A}_3$ is the 12×12 matrix in Eq. (3.35). This in turn leads to the following equations:

$$\bar{\mathbf{c}}_3 = \bar{\mathbf{A}}_3^{-1} \bar{\mathbf{u}} \qquad (3.36)$$

and

$$\begin{bmatrix} u \\ v \\ w \end{bmatrix} = \bar{\mathbf{C}}_3 \bar{\mathbf{A}}_3^{-1} \bar{\mathbf{u}} = \bar{\mathbf{a}}_3 \bar{\mathbf{u}} \qquad (3.37)$$

where

$$\bar{\mathbf{a}}_3 = \bar{\mathbf{C}}_3 \bar{\mathbf{A}}_3^{-1} \tag{3.38}$$

The mass matrix $\mathbf{m}_3$ is then given by

$$\mathbf{m}_3 = \rho \int_v \bar{\mathbf{a}}_3^T \bar{\mathbf{a}}_3 \, dV = \rho t \int_0^a \int_0^b \bar{\mathbf{a}}_3^T \bar{\mathbf{a}}_3 \, dx \, dy = \rho abt \int_0^1 \int_0^1 \bar{\mathbf{a}}_3^T \bar{\mathbf{a}}_3 \, d\xi \, d\eta$$

$$= \frac{\rho abt}{36}
\begin{bmatrix}
4 & 0 & 0 & 2 & 0 & 0 & 1 & 0 & 0 & 2 & 0 & 0 \\
0 & 4 & 0 & 0 & 2 & 0 & 0 & 1 & 0 & 0 & 2 & 0 \\
0 & 0 & 4 & 0 & 0 & 2 & 0 & 0 & 1 & 0 & 0 & 2 \\
2 & 0 & 0 & 4 & 0 & 0 & 2 & 0 & 0 & 1 & 0 & 0 \\
0 & 2 & 0 & 0 & 4 & 0 & 0 & 2 & 0 & 0 & 1 & 0 \\
0 & 0 & 2 & 0 & 0 & 4 & 0 & 0 & 2 & 0 & 0 & 1 \\
1 & 0 & 0 & 2 & 0 & 0 & 4 & 0 & 0 & 2 & 0 & 0 \\
0 & 1 & 0 & 0 & 2 & 0 & 0 & 4 & 0 & 0 & 2 & 0 \\
0 & 0 & 1 & 0 & 0 & 2 & 0 & 0 & 4 & 0 & 0 & 2 \\
2 & 0 & 0 & 1 & 0 & 0 & 2 & 0 & 0 & 4 & 0 & 0 \\
0 & 2 & 0 & 0 & 1 & 0 & 0 & 2 & 0 & 0 & 4 & 0 \\
0 & 0 & 2 & 0 & 0 & 1 & 0 & 0 & 2 & 0 & 0 & 4
\end{bmatrix} \tag{3.39}$$

6

Rectangular Plate T4: Assumed Stress Distribution

The node displacements for rectangular elements and the node numbering are shown in Fig. 4.1. The assumed stress field that satisfies the equations of stress equilibrium can be taken as

$$\sigma_{xx} = a_1 + a_2 y \tag{4.1}$$

$$\sigma_{yy} = a_3 + a_4 x \tag{4.2}$$

$$\sigma_{xy} = a_5 \tag{4.3}$$

where $a_1 \cdots a_5$ are constants. The assumed stress field and the resulting stiffness derivation are based on known concepts.

From the strain-stress equation for e_{xx} and the assumed stress field, it follows that

$$e_{xx} = \frac{\partial u}{\partial x} = \frac{1}{E}(\sigma_{xx} - v\sigma_{yy}) = \frac{1}{E}(a_1 + a_2 y - va_3 - va_4 x) \tag{4.4}$$

where E is the Young's modulus and v is the Poisson's ratio. Integration of this equation leads to

$$u = \frac{1}{E}\left(\frac{a_1 x + a_2 xy - va_3 x - va_4 x^2}{2}\right) + \frac{1}{E}f(y) \tag{4.5}$$

where $f(y)$ is an arbitrary function of y. Similarly, starting with the strain e_{yy}

$$e_{yy} = \frac{\partial v}{\partial y} = \frac{1}{E}(\sigma_{yy} - v\sigma_{xx}) = \frac{1}{E}(a_3 + a_4 x - va_1 - va_2 y) \tag{4.6}$$

from which it follows that

$$v = \frac{1}{E}\left(\frac{a_3 y + a_4 xy - va_1 y - va_2 y^2}{2}\right) + \frac{1}{E}g(x) \tag{4.7}$$

where $g(x)$ is an arbitrary function of x. Also, from the shearing stress equation

$$e_{xy} = \frac{\partial u}{\partial y} + \frac{\partial v}{\partial x} = \frac{\sigma_{xy}}{G} = \frac{1}{E}2(1 + v)a_5 \tag{4.8}$$

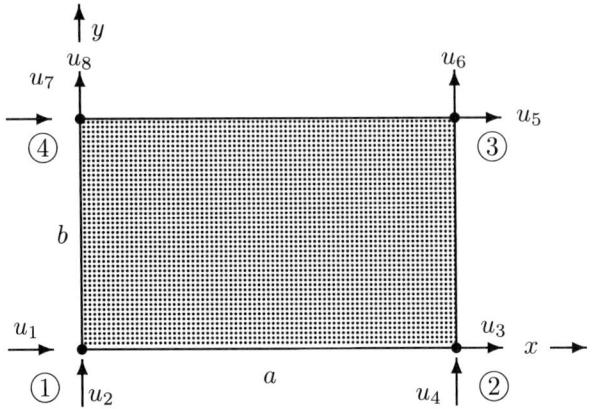

Fig. 4.1 Node displacements for rectangular plate elements.

[Note here that the shear modulus $G = E/2(1 + v)$.] To avoid the derivation of the coefficients $a_1 \cdots a_8$, we will use the results of [1–3] for the strain-node displacement relationship as shown here:

$$
\mathbf{e} = \begin{bmatrix} e_{xx} \\ e_{yy} \\ e_{xy} \end{bmatrix} = \begin{bmatrix} \partial u/\partial x \\ \partial v/\partial y \\ \partial u/\partial y + \partial v/\partial x \end{bmatrix} = \mathbf{b}_4 \mathbf{u} \tag{4.9}
$$

where

$$
\mathbf{b}_4 = \begin{bmatrix}
\dfrac{-(b-y)}{ab} & \dfrac{v(a-2x)}{2ab} & \dfrac{(b-y)}{ab} & \dfrac{-v(a-2x)}{2ab} & \dfrac{y}{ab} \\[2ex]
\dfrac{v(b-2y)}{2ab} & \dfrac{-(a-x)}{ab} & \dfrac{-v(b-2y)}{2ab} & \dfrac{-x}{ab} & \dfrac{v(b-2y)}{2ab} \\[2ex]
\dfrac{-1}{2b} & \dfrac{-1}{2a} & \dfrac{-1}{2b} & \dfrac{1}{2a} & \dfrac{1}{2b}
\end{bmatrix}
$$

$$
\begin{matrix}
\dfrac{v(a-2x)}{2ab} & \dfrac{-y}{ab} & \dfrac{-v(a-2x)}{2ab} \\[2ex]
\dfrac{x}{ab} & \dfrac{-v(b-2y)}{2ab} & \dfrac{(a-x)}{ab} \\[2ex]
\dfrac{1}{2a} & \dfrac{1}{2b} & \dfrac{-1}{2a}
\end{matrix} \tag{4.10}
$$

and

$$
\mathbf{u} = \begin{bmatrix} u_1 \\ \vdots \\ u_8 \end{bmatrix} \tag{4.11}
$$

The stresses within the element are calculated from the stress-strain equations for the plane stress field and are given by

$$\begin{bmatrix} \sigma_{xx} \\ \sigma_{yy} \\ \sigma_{xy} \end{bmatrix} = \frac{E}{(1-v^2)} \begin{bmatrix} 1 & v & 0 \\ v & 1 & 0 \\ 0 & 0 & (1-v)/2 \end{bmatrix} \begin{bmatrix} e_{xx} \\ e_{yy} \\ e_{xy} \end{bmatrix}$$
$$- \frac{\alpha T}{(1-v)} \begin{bmatrix} 1 \\ 1 \\ 0 \end{bmatrix} \tag{4.12}$$

which can be expressed symbolically as

$$\sigma = E_2 e - \frac{\alpha T}{(1-v)} \begin{bmatrix} 1 \\ 1 \\ 0 \end{bmatrix} \tag{4.13}$$

where

$$E_2 = \frac{E}{(1-v^2)} \begin{bmatrix} 1 & v & 0 \\ v & 1 & 0 \\ 0 & 0 & (1-v)/2 \end{bmatrix} \tag{4.14}$$

The stiffness matrix for the rectangular plate element is then obtained from

$$k_4 = \int_V b_4^T E_2 b_4 \, dV$$
$$= \int_0^b \int_0^a b_4^T E_2 b_4 \, dx \, dy = abt \int_0^1 \int_0^1 b_4^T E_2 b_4 \, d\xi \, d\eta \tag{4.15}$$

where t is the plate thickness (assumed constant) and

$$\xi = \frac{x}{a} \tag{4.16}$$

$$\eta = \frac{y}{b} \tag{4.17}$$

and

$$\beta = \frac{b}{a} \tag{4.18}$$

It can therefore be shown that the stiffness matrix k_4 for a rectangular plate derived from the assumed linear distribution of stresses is given by

$$k_4 = \begin{bmatrix} k_{11} & k_{12} \\ k_{21} & k_{22} \end{bmatrix} \tag{4.19}$$

where

$$\mathbf{k}_{11} = \frac{Et}{12(1-v^2)} \begin{bmatrix} (4-v^2)\beta & \frac{3}{2}(1+v) & -(4-v^2)/\beta & -\frac{3}{2}(1-3v) \\ +\frac{3}{2}(1-v)/\beta & & +\frac{3}{2}(1-v)\beta & \\ \frac{3}{2}(1+v) & (4-v^2)/\beta & \frac{3}{2}(1-3v) & (2+v^2)/\beta \\ & +\frac{3}{2}(1-v)\beta & -\frac{3}{2}(1-v)\beta & \\ -(4-v^2)\beta & \frac{3}{2}(1-3v) & (4-v^2)\beta & -\frac{3}{2}(1+v) \\ +\frac{3}{2}(1-v)/\beta & & +\frac{3}{2}(1-v)/\beta & +\frac{3}{2}(1-v)\beta \\ -\frac{3}{2}(1-3v) & (2+v^2)\beta & -\frac{3}{2}(1+v) & (4-v^2)/\beta \\ & -\frac{3}{2}(1-v)\beta & & \frac{3}{2}(1-v)\beta \end{bmatrix}$$

(4.20a)

$$\mathbf{k}_{21} = \frac{Et}{12(1-v^2)} \begin{bmatrix} -(2+v^2)\beta & -\frac{3}{2}(1+v) & (2+v^2)\beta & \frac{3}{2}(1-3v) \\ -\frac{3}{2}(1-v)\beta & & -\frac{3}{2}(1-v)/\beta & \\ -\frac{3}{2}(1+v) & -(2+v^2)/\beta & -\frac{3}{2}(1-3v) & -(4-v^2)\beta \\ & -\frac{3}{2}(1-v)\beta & & +\frac{3}{2}(1-v)\beta \\ (2+v^2)\beta & -\frac{3}{2}(1-3v) & -(2-v^2)\beta & \frac{3}{2}(1+v) \\ -\frac{3}{2}(1-v)/\beta & & -\frac{3}{2}(1-v)/\beta & \\ \frac{3}{2}(1-3v) & -(4-v^2)/\beta & \frac{3}{2}(1+v) & -(2+v^2)/\beta \\ & +\frac{3}{2}(1-v)\beta & & -\frac{3}{2}(1-v)\beta \end{bmatrix}$$

(4.20b)

$$\mathbf{k}_{12} = \mathbf{k}_{21}^T$$

(4.20c)

$$\mathbf{k}_{22} = \frac{Et}{12(1-v^2)} \begin{bmatrix} (4-v^2)\beta & \frac{3}{2}(1+v) & -(4-v^2)\beta & -\frac{3}{2}(1-3v) \\ +\frac{3}{2}(1-v)/\beta & & +\frac{3}{2}(1-v)/\beta & \\ \frac{3}{2}(1+v) & (4-v^2)/\beta & \frac{3}{2}(1-3v) & (2+v^2)/\beta \\ & +\frac{3}{2}(1-v)\beta & & -\frac{3}{2}(1-v)\beta \\ -(4-v^2)\beta & \frac{3}{2}(1-3v) & (4-v^2)\beta & -\frac{3}{2}(1+v) \\ +\frac{3}{2}(1-v)/\beta & & +\frac{3}{2}(1-v)/\beta & \\ -\frac{3}{2}(1-3v) & (2+v^2)/\beta & -\frac{3}{2}(1+v) & (4-v^2)/\beta \\ & -\frac{3}{2}(1-v)\beta & & +\frac{3}{2}(1-v)\beta \end{bmatrix}$$

$$(4.20d)$$

and where

$$\beta = \frac{a}{b} \tag{4.21}$$

For $\beta = 1$ and $v = 0.3$, the stiffness matrix $\mathbf{k}_4$ based on the linear stress assumption is given by

$$\mathbf{k}_4 = Et$$

$$\times \begin{bmatrix} 0.4542 & 0.1786 & -0.2619 & -0.0137 \\ 0.1786 & 0.4542 & 0.0137 & 0.0952 \\ -0.2619 & 0.0137 & 0.4542 & -0.1786 \\ -0.0137 & 0.0952 & -0.1786 & 0.4542 \\ -0.2875 & -0.1786 & 0.0952 & 0.0137 \\ -0.1786 & -0.2875 & -0.0137 & -0.2619 \\ 0.0952 & -0.0137 & -0.2875 & 0.1786 \\ 0.0137 & -0.2619 & 0.1786 & -0.2875 \end{bmatrix}$$

$$\begin{bmatrix} -0.2875 & -0.1786 & 0.0952 & 0.0137 \\ -0.1786 & -0.2875 & -0.0137 & -0.2619 \\ 0.0952 & -0.0137 & -0.2875 & 0.1786 \\ 0.0137 & -0.2619 & 0.1786 & -0.2875 \\ 0.4542 & 0.1786 & -0.2619 & -0.0137 \\ 0.1786 & 0.4542 & 0.0137 & 0.0952 \\ -0.2619 & 0.0137 & 0.4542 & -0.1786 \\ -0.0137 & 0.0952 & -0.1786 & 0.4542 \end{bmatrix} \quad (4.22)$$

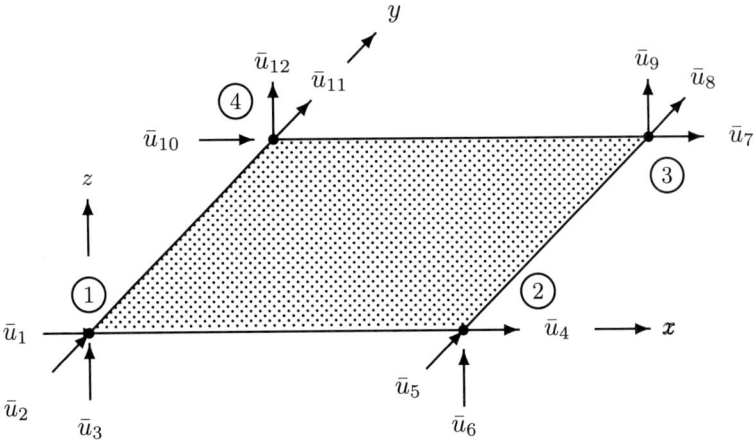

Fig. 4.2   Node displacements for the mass matrix m₄.

Comparison of the diagonal stiffness coefficients for T4 and T3 indicates that the ratio of these coeficients is 0.92, that is, T4 element is less stiff than T3.

Thermal Stiffness, Thermal Loads, and Mass Matrices

As an approximation, the thermal stiffness, thermal load, and mass matrices can be assumed to be the same as for the T3 element, that is,

$$\mathbf{h}_4 \approx \mathbf{h}_3 \tag{4.23}$$

$$\mathbf{q}_4 \approx \mathbf{q}_3 \tag{4.24}$$

$$\mathbf{m}_4 \approx \mathbf{m}_3 \tag{4.25}$$

The node displacements for the mass matrix $\mathbf{m}_4$ are shown in Fig. 4.2.

7

Quadrilateral Plate T5: Assumed Displacement Distribution

The node displacements for quadrilateral plate elements and the node numbering are shown in Fig. 5.1.

The assumed displacement field for the quadrilateral element can be taken as

$$u = c_1 + c_2 x + c_3 xy + c_4 y \tag{5.1}$$

$$v = c_5 + c_6 x + c_7 xy + c_8 y \tag{5.2}$$

where u and v are the displacements in the x and y directions as shown in Fig. 5.1. Hence, the element node displacements $\mathbf{u}$ can be expressed as

$$\mathbf{u} = \begin{bmatrix} u_1 \\ u_2 \\ u_3 \\ u_4 \\ u_5 \\ u_6 \\ u_7 \\ u_8 \end{bmatrix} = \begin{bmatrix} 1 & x_1 & x_1 y_1 & y_1 & 0 & 0 & 0 & 0 \\ 0 & 0 & 0 & 0 & 1 & x_1 & x_1 y_1 & y_1 \\ 1 & x_2 & x_2 y_2 & y_2 & 0 & 0 & 0 & 0 \\ 0 & 0 & 0 & 0 & 1 & x_2 & x_2 y_2 & y_2 \\ 1 & x_3 & x_3 y_3 & y_3 & 0 & 0 & 0 & 0 \\ 0 & 0 & 0 & 0 & 1 & x_3 & x_3 y_3 & y_3 \\ 1 & x_4 & x_4 y_4 & y_4 & 0 & 0 & 0 & 0 \\ 0 & 0 & 0 & 0 & 1 & x_4 & x_4 y_4 & y_4 \end{bmatrix} \begin{bmatrix} c_1 \\ c_2 \\ c_3 \\ c_4 \\ c_5 \\ c_6 \\ c_7 \\ c_8 \end{bmatrix} \tag{5.3}$$

Symbolically, the preceding equation is expressed as

$$\mathbf{u} = \mathbf{C}_5 \mathbf{c}_5 \tag{5.4}$$

from which

$$\mathbf{c}_5 = \mathbf{C}_5^{-1} \mathbf{u} \tag{5.5}$$

The strains $\mathbf{e}$ in the element are then computed from the assumed displacements as

$$\mathbf{e} = \begin{bmatrix} \partial u/\partial x \\ \partial v/\partial y \\ \partial u/\partial y + \partial v/\partial x \end{bmatrix} = \begin{bmatrix} 0 & 1 & y & 0 & 0 & 0 & 0 & 0 \\ 0 & 0 & 0 & 0 & 0 & 0 & x & 1 \\ 0 & 0 & x & 1 & 0 & 1 & y & 0 \end{bmatrix} \begin{bmatrix} c_1 \\ c_2 \\ \vdots \\ c_7 \\ c_8 \end{bmatrix} \tag{5.6}$$

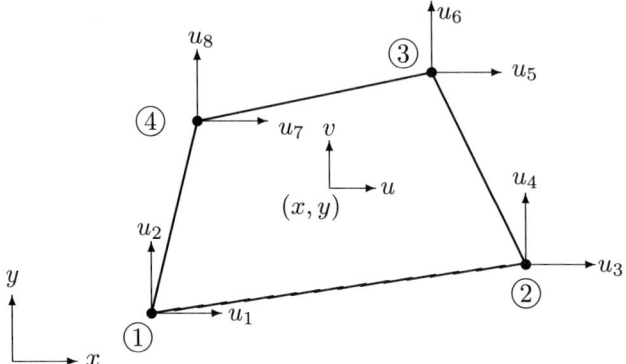

Fig. 5.1   Node displacements for quadrilateral plate elements.

or symbolically as

$$\mathbf{e} = \mathbf{A}_5\mathbf{c} = \mathbf{A}_5\mathbf{C}_5^{-1}\mathbf{u} = \mathbf{b}_5\mathbf{u} \tag{5.7}$$

where $\mathbf{A}_5$ is the preceding (3×8) matrix and

$$\mathbf{b}_5 = \mathbf{A}_5\mathbf{C}_5^{-1} \tag{5.8}$$

Here the matrix $\mathbf{b}_5$ represents a matrix of strains caused by unit displacements at the element nodes. These strains satisfy equations of strain compatibility within the element, but they do violate equations of stress equilibrium.

The stress-strain equations for the plane stress field are given by

$$
\boldsymbol{\sigma} =
\begin{bmatrix} \sigma_{xx} \\ \sigma_{yy} \\ \sigma_{xy} \end{bmatrix}
= \frac{E}{(1 - \nu^2)}
\begin{bmatrix} 1 & \nu & 0 \\ \nu & 1 & 0 \\ 0 & 0 & (1 - \nu)/2 \end{bmatrix}
\begin{bmatrix} e_{xx} \\ e_{yy} \\ e_{xy} \end{bmatrix}
- \frac{\alpha T}{(1 - \nu)}
\begin{bmatrix} 1 \\ 1 \\ 0 \end{bmatrix}
$$

$$
= \mathbf{E}_2\mathbf{e} - \frac{\alpha T}{(1 - \nu)}
\begin{bmatrix} 1 \\ 1 \\ 0 \end{bmatrix} \tag{5.9}
$$

where

$$
\mathbf{E}_2 = \frac{E}{(1 - \nu^2)}
\begin{bmatrix} 1 & \nu & 0 \\ \nu & 1 & 0 \\ 0 & 0 & (1 - \nu)/2 \end{bmatrix} \tag{5.10}
$$

To calculate the element stiffness matrix $\mathbf{k}_5$, the natural coordinate system (ξ, η) is introduced as shown in Fig. 5.2. This system allows for a simple numerical integration over the quadrilateral.

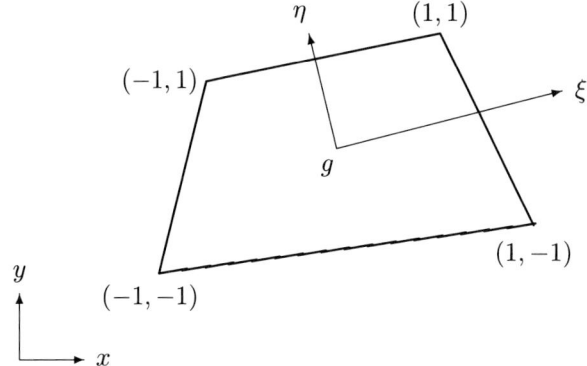

Fig. 5.2   Natural coordinate system for quadrilateral elements.

The geometric center g for the quadrilateral coordinate system is defined as

$$x_g = \frac{(x_1 + x_2 + x_3 + x_4)}{4} \tag{5.11}$$

$$y_g = \frac{(y_1 + y_2 + y_3 + y_4)}{4} \tag{5.12}$$

The rectangular coordinates x and y can be expressed in terms of the natural coordinates ξ and η as

$$x = \sum_{i=1}^{4} f_i x_i \tag{5.13}$$

and

$$y = \sum_{i=1}^{4} f_i y_i \tag{5.14}$$

where

$$f_1 = \frac{(1 - \xi)(1 - \eta)}{4} \tag{5.15a}$$

$$f_2 = \frac{(1 + \xi)(1 - \eta)}{4} \tag{5.15b}$$

$$f_3 = \frac{(1 + \xi)(1 + \eta)}{4} \tag{5.15c}$$

$$f_4 = \frac{(1 - \xi)(1 + \eta)}{4} \tag{5.15d}$$

The stiffness matrix for the quadrilateral element is obtained from

$$\mathbf{k}_5 = \int_V \mathbf{b}_5^T \mathbf{E}_2 \mathbf{b}_5 \, dV = t \iint \mathbf{b}_5^T \mathbf{E}_2 \mathbf{b}_5 \, dx \, dy = t \iint \mathbf{b}_5^T \mathbf{E}_2 \mathbf{b}_5 |\mathbf{J}(\xi, \eta)| \, d\xi \, d\eta$$

(5.16)

where t is the panel thickness (assumed to be constant) and $|\mathbf{J}|$ is the determinant of the Jacobian $\mathbf{J}$, which is expressed as

$$\mathbf{J}(\xi, \eta) = \begin{bmatrix} J_{11} & J_{12} \\ J_{21} & J_{22} \end{bmatrix} = \begin{bmatrix} x, \xi & y, \xi \\ x, \eta & y, \eta \end{bmatrix}$$

(5.17)

Here the subscripts ξ and η denote differentiation with respect to ξ and η of the previously defined expressions for x and y in terms of the natural coordinates ξ and η. The integration of the expression for $\mathbf{k}_5$ can be performed numerically using the Gaussian quadrature method.

Thermal Stiffness

The thermal stiffness for this element is computed from

$$\mathbf{h}_5 = \int_v \mathbf{b}_5^T \frac{E}{(1-v)} \begin{bmatrix} -1 \\ -1 \\ 0 \end{bmatrix} dV = \frac{Et}{(1-v)} \int_{-1}^1 \int_{-1}^1 \mathbf{b}_5^T \begin{bmatrix} -1 \\ -1 \\ 0 \end{bmatrix} |J(\xi, \eta)| \, d\xi \, d\eta$$

(5.18)

Thermal Load

The thermal load for the element is obtained from

$$\mathbf{q}_5 = \alpha T \mathbf{h}_5$$

(5.19)

where T is the temperature of the element (assumed to be constant) and α is the coefficients of thermal expansion.

Mass Matrix

The node displacement for use in the element mass matrix is shown in Fig. 5.3. In addition to the displacements u and v, the transverse displacement in the z direction will be assumed to be of the same form as for u and v, that is,

$$w = c_9 + c_{10}x + c_{11}xy + c_{12}y$$

(5.20)

which, when combined with Eqs. (5.1) and (5.2), leads to

$$\begin{bmatrix} u \\ v \\ w \end{bmatrix} = \begin{bmatrix} 1 & x & xy & y & 0 & 0 & 0 & 0 & 0 & 0 & 0 & 0 \\ 0 & 0 & 0 & 0 & 1 & x & xy & y & 0 & 0 & 0 & 0 \\ 0 & 0 & 0 & 0 & 0 & 0 & 0 & 0 & 1 & x & xy & y \end{bmatrix} \begin{bmatrix} c_1 \\ \vdots \\ c_{12} \end{bmatrix}$$

$$= \bar{\mathbf{C}}_5 \bar{\mathbf{c}}_5$$

(5.21)

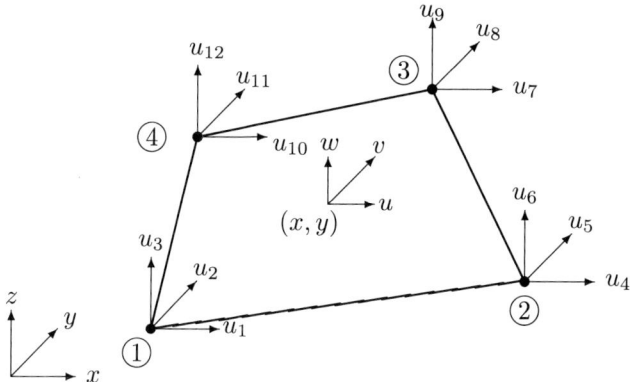

Fig. 5.3   Node displacements of a quadrilateral panel used for the mass matrix.

where $\bar{\mathbf{C}}_5$ is the preceding (3×12) matrix and

$$\bar{\mathbf{c}}_5 = \{c_1 \cdots c_{12}\} \tag{5.22}$$

Hence, from Eq. (5.21)

$$\bar{\mathbf{u}} = \begin{bmatrix} u_1 \\ u_2 \\ u_3 \\ u_4 \\ u_5 \\ u_6 \\ u_7 \\ u_8 \\ u_9 \\ u_{10} \\ u_{11} \\ u_{12} \end{bmatrix} = \begin{bmatrix} 1 & x_1 & x_1y_1 & y_1 & 0 & 0 & 0 & 0 & 0 & 0 & 0 & 0 \\ 0 & 0 & 0 & 0 & 1 & x_1 & x_1y_1 & y_1 & 0 & 0 & 0 & 0 \\ 0 & 0 & 0 & 0 & 0 & 0 & 0 & 0 & 1 & x_1 & x_1y_1 & y_1 \\ 1 & x_2 & x_2y_2 & y_2 & 0 & 0 & 0 & 0 & 0 & 0 & 0 & 0 \\ 0 & 0 & 0 & 0 & 1 & x_2 & x_2y_2 & y_2 & 0 & 0 & 0 & 0 \\ 0 & 0 & 0 & 0 & 0 & 0 & 0 & 0 & 1 & x_2 & x_2y_2 & y_2 \\ 1 & x_3 & x_3y_3 & y_3 & 0 & 0 & 0 & 0 & 0 & 0 & 0 & 0 \\ 0 & 0 & 0 & 0 & 1 & x_3 & x_3y_3 & y_3 & 0 & 0 & 0 & 0 \\ 0 & 0 & 0 & 0 & 0 & 0 & 0 & 0 & 1 & x_3 & x_3y_3 & y_3 \\ 1 & x_4 & x_4y_4 & y_4 & 0 & 0 & 0 & 0 & 0 & 0 & 0 & 0 \\ 0 & 0 & 0 & 0 & 1 & x_4 & x_4y_4 & 0 & 0 & 0 & 0 & 0 \\ 0 & 0 & 0 & 0 & 0 & 0 & 0 & y_4 & 1 & y_4 & x_4y_4 & y_4 \end{bmatrix} \begin{bmatrix} c_1 \\ c_2 \\ c_3 \\ c_4 \\ c_5 \\ c_6 \\ c_7 \\ c_8 \\ c_9 \\ c_{10} \\ c_{11} \\ c_{12} \end{bmatrix}$$

$$= \bar{\mathbf{A}}_5 \bar{\mathbf{c}}_5 \tag{5.23}$$

where $\bar{\mathbf{A}}_5$ is the (12×12) matrix in Eq. (5.21). Therefore,

$$\bar{\mathbf{c}}_5 = \bar{\mathbf{A}}_5^{-1} \bar{\mathbf{u}} \tag{5.24}$$

and

$$\begin{bmatrix} u \\ v \\ w \end{bmatrix} = \bar{\mathbf{C}}_5 \bar{\mathbf{A}}_5^{-1} \bar{\mathbf{u}} = \bar{\mathbf{a}}_5 \bar{\mathbf{u}} \tag{5.25}$$

where

$$\bar{\mathbf{a}}_5 = \bar{\mathbf{C}}_5 \bar{\mathbf{A}}_5^{-1} \tag{5.26}$$

is the shape function and

$$\bar{\mathbf{u}} = \{u_1 \cdots u_{12}\} \tag{5.27}$$

The mass matrix $\mathbf{m}_5$ for this element is then determined from

$$\mathbf{m}_5 = \rho \iiint \bar{\mathbf{a}}_5^T \bar{\mathbf{a}}_5 \, dx \, dy \, dz = \rho t \int_{-1}^{1} \int_{-1}^{1} \bar{\mathbf{a}}_5^T \bar{\mathbf{a}}_5 |\mathbf{J}(\xi, \eta)| \, d\xi \, d\eta \tag{5.28}$$

where $|\mathbf{J}(x, y)|$ is the Jacobian introduced earlier and the integration is performed using the Gaussian quadrature.

8

Quadrilateral Plate T6: Assumed Stress Distribution

The node displacements for quadrilateral elements and the node numbering are shown in Fig. 6.1.

The assumed stress field for this element is

$$\sigma_{xx} = a_1 + a_2 y \tag{6.1}$$

$$\sigma_{yy} = a_3 + a_4 x \tag{6.2}$$

$$\sigma_{xy} = a_5 \tag{6.3}$$

From the strain-stress equation for e_{xx} and the assumed stress field, it follows that

$$e_{xx} = \frac{\partial u}{\partial x} = \frac{1}{E}(\sigma_{xx} - v\sigma_{yy}) = \frac{1}{E}(a_1 + a_2 y - va_3 - va_4 x) \tag{6.4}$$

where E is the Young's modulus and v is the Poisson's ratio. Integration of this equation leads to

$$u = \frac{1}{E}\left(\frac{a_1 x + a_2 xy - va_3 x - va_4 x^2}{2}\right) + \frac{1}{E}f(y) \tag{6.5}$$

where $f(y)$ is an arbitrary function of y. Similarly, starting with the strain e_{yy}

$$e_{yy} = \frac{\partial v}{\partial y} = \frac{1}{E}(\sigma_{yy} - v\sigma_{xx}) = \frac{1}{E}(a_3 + a_4 x - va_1 - va_2 y) \tag{6.6}$$

from which it follows that

$$v = \frac{1}{E}\left(\frac{a_3 y + a_4 xy - va_1 y - va_2 y^2}{2}\right) + \frac{1}{E}g(x) \tag{6.7}$$

where $g(x)$ is an arbitrary function of x. Also, from the shearing stress equation

$$e_{xy} = \frac{\partial u}{\partial y} + \frac{\partial v}{\partial x} = \frac{\sigma_{xy}}{G} = \frac{1}{E}2(1 + v)a_5 \tag{6.8}$$

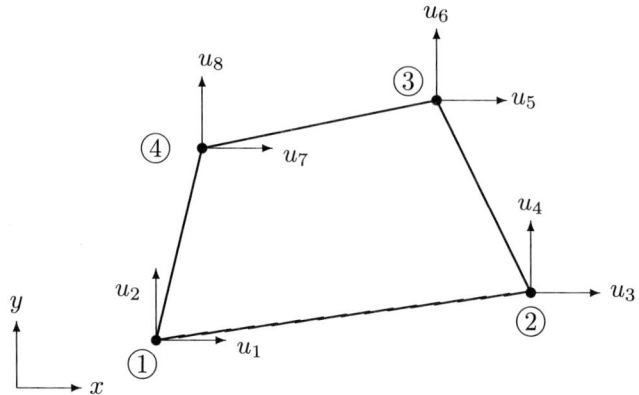

Fig. 6.1   Node displacements for quadrilateral panel elements.

[Note here that the shear modulus $G = E/2(1 + v)$.] Substituting now the derivatives of u and v into the shear equation and multiplying it throughout by E, it follows that

$$a_2x + f'(y) + a_4y + g'(x) = 2(1 + v)a_5 \qquad (6.9)$$

which after rearranging becomes

$$f'(y) + a_4y = 2(1 + v)a_5 - g'(x) - a_2x = a_6 \qquad (6.10)$$

where a_6 represents a constant and primes denote derivatives with respect to the indicated variables. The preceding equation can only be satisfied if both sides are equal to the constant a_6. Rearranging Eq. (6.10), it follows that

$$f'(y) = a_6 - a_4y \qquad (6.11)$$

and

$$g'(x) = 2(1 + v)a_5 - a_2x - a_6 \qquad (6.12)$$

Hence solving for $f(y)$ and $g(x)$,

$$f(y) = a_6y - a_4y^2/2 + a_7 \qquad (6.13)$$

and

$$g(x) = [2(1 + v)a_5 - a_6]x - a_2x^2/2 + a_8 \qquad (6.14)$$

The constants of integration a_7 and a_8 represent the rigid-body translations, while the previously introduced constant a_6 defines the rigid-body rotation.

Substitution of the expressions for $f(y)$ and $g(x)$ into the expressions for u and v [Eqs. (6.5) and (6.7)] leads to the following two equations:

$$u = \frac{1}{E}\left[a_1 x + a_2 xy - \nu a_3 x - a_4(\nu x^2 + y^2)/2 + a_6 y + a_7\right]$$
$$= c_1 x + c_2 y - c_3(\nu x^2 + y^2) + 2c_4 xy + c_5 \tag{6.15}$$

and

$$v = \frac{1}{E}\left[-\nu a_1 y - a_2(x^2 + \nu y^2)/2 + a_3 y + a_4 xy + 2(1 + \nu)a_5 x - a_6 x + a_8\right]$$
$$= 2c_3 xy - c_4(x^2 + \nu y^2) + c_6 x + c_7 y + c_8 \tag{6.16}$$

where

$$c_1 = (a_1 - \nu a_3)/E \tag{6.17a}$$
$$c_2 = a_6/E \tag{6.17b}$$
$$c_3 = a_4/2E \tag{6.17c}$$
$$c_4 = a_2/E \tag{6.17d}$$
$$c_5 = a_7/E \tag{6.17e}$$
$$c_6 = [2(1 + \nu)a_5 - a_6]/E \tag{6.17f}$$
$$c_7 = (a_3 - \nu a_1)/E \tag{6.17g}$$
$$c_8 = a_8/E \tag{6.17h}$$

The unknown constants $c_1 \cdots c_8$ can now be determined from the element displacements $u_1 \cdots u_8$. Hence using Eqs. (6.15) and (6.16),

$$\mathbf{u} = \begin{bmatrix} u_1 \\ u_2 \\ u_3 \\ u_4 \\ u_5 \\ u_6 \\ u_7 \\ u_8 \end{bmatrix}$$

$$= \begin{bmatrix} x1 & y1 & -(\nu x_1^2 + y_1^2) & 2x_1 y_1 & 1 & 0 & 0 & 0 \\ 0 & 0 & 2x_1 y_1 & (x_1^2 + \nu y_1^2) & 0 & x1 & y1 & 1 \\ x2 & y2 & -(\nu x_2^2 + y_2^2) & 2x_2 y_2 & 1 & 0 & 0 & 0 \\ 0 & 0 & 2x_2 y_2 & (x_2^2 + \nu y_2^2) & 0 & x2 & y2 & 1 \\ x3 & y3 & -(\nu x_3^2 + y_3^2) & 2x_3 y_3 & 1 & 0 & 0 & 0 \\ 0 & 0 & 2x_3 y_3 & (x_3^2 + \nu y_3^2) & 0 & x3 & y3 & 1 \\ x4 & y4 & -(\nu x_4^2 + y_4^2) & 2x_4 y_4 & 1 & 0 & 0 & 0 \\ 0 & 0 & 2x_4 y_{41} & (x_4^2 + \nu y_4^2) & 0 & x4 & y4 & 1 \end{bmatrix} \begin{bmatrix} c_1 \\ c_2 \\ c_3 \\ c_4 \\ c_5 \\ c_6 \\ c_7 \\ c_8 \end{bmatrix} \tag{6.18}$$

Symbolically the preceding equation can be expressed as

$$\mathbf{u} = \mathbf{C}_6 \mathbf{c}_6 \tag{6.19}$$

or as

$$\mathbf{c}_6 = \mathbf{C}_6^{-1} \mathbf{u} \tag{6.20}$$

The strains in the element are then computed from

$$\mathbf{e} = \begin{bmatrix} \partial u/\partial x \\ \partial v/\partial y \\ \partial u/\partial y + \partial v/\partial x \end{bmatrix}$$

$$= \begin{bmatrix} 1 & 0 & -2vx & 2y & -2vx & 0 & 0 & 0 \\ 0 & 0 & 2x & 2x & -2vy & 0 & 1 & 0 \\ 0 & 1 & 0 & -2(x+y) & -2(x+y) & 1 & 0 & 0 \end{bmatrix} \begin{bmatrix} c_1 \\ c_2 \\ \vdots \\ c_7 \\ c_8 \end{bmatrix} \tag{6.21}$$

Hence

$$\mathbf{e} = \mathbf{A}_6 \mathbf{c}_6 = \mathbf{A}_6 \mathbf{C}_6^{-1} \mathbf{u} = \mathbf{b}_6 \mathbf{u} \tag{6.22}$$

and

$$\mathbf{b}_6 = \mathbf{A}_6 \mathbf{C}_6^{-1} \tag{6.23}$$

where $\mathbf{A}_6$ is the (3×8) matrix above with the factor $1/E$ and $\mathbf{b}_6$ represents a matrix of strains caused by unit displacements at the element nodes. These strains satisfy both equations of strain compatibility and stress equilibrium. The expression for $\mathbf{b}_6$ is shown next:

$$\mathbf{b}_6 = \begin{bmatrix} \dfrac{-(1-\eta)}{a} & \dfrac{v(1-2\xi)}{2b} & \dfrac{(1-\eta)}{a} & \dfrac{-v(1-2\xi)}{2b} \\[2mm] \dfrac{v(1-2\eta)}{2a} & \dfrac{-(1-\xi)}{b} & \dfrac{-v(1-2\eta)}{2a} & \dfrac{-\xi}{b} \\[2mm] \dfrac{-1}{2b} & \dfrac{-1}{2a} & \dfrac{-1}{2b} & \dfrac{1}{2a} \end{bmatrix}$$

$$\begin{matrix} \dfrac{\eta}{a} & \dfrac{v(1-2\xi)}{2b} & \dfrac{-\eta}{a} & \dfrac{-v(1-2\xi)}{2b} \\[2mm] \dfrac{v(1-2\eta)}{2a} & \dfrac{\xi}{b} & \dfrac{-(1-2\eta)}{2a} & \dfrac{(1-\xi)}{b} \\[2mm] \dfrac{1}{2b} & \dfrac{1}{2a} & \dfrac{1}{2b} & \dfrac{-1}{2a} \end{matrix} \tag{6.24}$$

The stresses within the element are calculated from the stress-strain equations for the plane stress field given by

$$\begin{bmatrix} \sigma_{xx} \\ \sigma_{yy} \\ \sigma_{xy} \end{bmatrix} = \frac{E}{(1-v^2)} \begin{bmatrix} 1 & v & 0 \\ v & 1 & 0 \\ 0 & 0 & (1-v)/2 \end{bmatrix} \begin{bmatrix} e_{xx} \\ e_{yy} \\ e_{xy} \end{bmatrix} - \frac{\alpha T}{(1-v)} \begin{bmatrix} 1 \\ 1 \\ 0 \end{bmatrix} \tag{6.25}$$

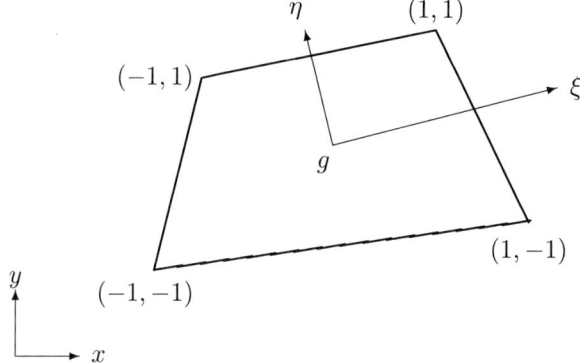

Fig. 6.2 Natural coordinate system for quadrilateral elements.

which can be expressed symbolically as

$$\boldsymbol{\sigma} = \mathbf{E}_2\mathbf{e} - \frac{\alpha T}{(1-\nu)}\begin{bmatrix}1\\1\\0\end{bmatrix} \tag{6.26}$$

where

$$\mathbf{E}_2 = \frac{E}{(1-\nu^2)}\begin{bmatrix}1 & \nu & 0\\ \nu & 1 & 0\\ 0 & 0 & (1-\nu)/2\end{bmatrix} \tag{6.27}$$

Introducing now the natural coordinate system (ξ, η), as shown in Fig. 6.2, and defining the geometric center g for the quadrilaterals as

$$x_g = \frac{x_1 + x_2 + x_3 + x_4}{4} \tag{6.28}$$

$$y_g = \frac{y_1 + y_2 + y_3 + y_4}{4} \tag{6.29}$$

the rectangular coordinates x and y can be expressed in terms of the natural coordinates ξ and η as

$$x = \sum_{i=1}^{4} f_i x_i \tag{6.30}$$

and

$$y = \sum_{i=1}^{4} f_i y_i \tag{6.31}$$

where

$$f_1 = \frac{(1 - \xi)(1 - \eta)}{4} \tag{6.32a}$$

$$f_2 = \frac{(1 + \xi)(1 - \eta)}{4} \tag{6.32b}$$

$$f_3 = \frac{(1 + \xi)(1 + \eta)}{4} \tag{6.32c}$$

$$f_4 = \frac{(1 - \xi)(1 + \eta)}{4} \tag{6.32d}$$

The introduction of the natural coordinate system allows the performance of the integration over the quadrilateral area of the element so that the stiffness matrix for the quadrilateral can be obtained from

$$\mathbf{k}_6 = \int_V \mathbf{b}_6^T \mathbf{E}_2 \mathbf{b}_6 \, dV = t \iint \mathbf{b}_6^T \mathbf{E}_2 \mathbf{b}_6 \, dx \, dy = t \int_{-1}^{1} \int_{-1}^{1} \mathbf{b}_6^T \mathbf{E}_2 \mathbf{b}_6 |\mathbf{J}(\xi, \eta)| \, d\xi \, d\eta \tag{6.33}$$

where t is the panel thickness (assumed constant) and $|\mathbf{J}|$ is the determinant of the Jacobian $\mathbf{J}$ that can be expressed as

$$\mathbf{J}(\xi, \eta) = \begin{bmatrix} J_{11} & J_{12} \\ J_{21} & J_{22} \end{bmatrix} = \begin{bmatrix} x_{,\xi} & y_{,\xi} \\ x_{,\eta} & y_{,\eta} \end{bmatrix} \tag{6.34}$$

in which the subscripts ξ and η denote differentiation with respect to ξ and η of the previously defined expressions for x and y in terms of the natural coordinates ξ and η. The integration of the expression for $\mathbf{k}_6$ can be performed numerically using the Gaussian quadrature method.

Thermal Stiffness

The thermal stiffness for this element is computed from

$$\mathbf{h}_6 = \int_v \mathbf{b}_6^T \frac{E}{(1 - v)} \begin{bmatrix} -1 \\ -1 \\ 0 \end{bmatrix} dV = \frac{Et}{(1 - v)} \int_{-1}^{1} \int_{-1}^{1} \mathbf{b}_6^T \begin{bmatrix} -1 \\ -1 \\ 0 \end{bmatrix} |J(\xi, \eta)| \, d\xi \, d\eta \tag{6.35}$$

Thermal Load

The thermal load for the element is obtained from

$$\mathbf{q}_6 = \alpha T \mathbf{h}_6 \tag{6.36}$$

where T is the temperature of the element (assumed to be constant) and α is the coefficients of thermal expansion.

Mass Matrix

For the mass matrix, three displacements in the x, y, and z directions will be used at each node point as it was done for the T5 element. The node displacement for use in the element mass matrix are shown in Fig. 6.3. In addition to the displacements u and v, the transverse displacement in the z direction will be assumed to be of the same form as for u and v, that is,

$$w = c_9 + c_{10}x + c_{11}xy + c_{12}y \tag{6.37}$$

which when combined with Eqs. (6.1) and (6.2) leads to

$$\begin{bmatrix} u \\ v \\ w \end{bmatrix} = \begin{bmatrix} 1 & x & xy & y & 0 & 0 & 0 & 0 & 0 & 0 & 0 & 0 \\ 0 & 0 & 0 & 0 & 1 & x & xy & y & 0 & 0 & 0 & 0 \\ 0 & 0 & 0 & 0 & 0 & 0 & 0 & 0 & 1 & x & xy & y \end{bmatrix} \begin{bmatrix} c_1 \\ \vdots \\ c_{12} \end{bmatrix} = \bar{C}_6 \bar{c}_6 \tag{6.38}$$

where $\bar{C}_5$ is the preceding (3×12) matrix and

$$\bar{c}_5 = \{c_1 \cdots c_{12}\} \tag{6.39}$$

Hence from Eq. (6.38)

$$\bar{u} = \begin{bmatrix} u_1 \\ u_2 \\ u_3 \\ u_4 \\ u_5 \\ u_6 \\ u_7 \\ u_8 \\ u_9 \\ u_{10} \\ u_{11} \\ u_{12} \end{bmatrix}$$

$$= \begin{bmatrix} 1 & x_1 & x_1y_1 & y_1 & 0 & 0 & 0 & 0 & 0 & 0 & 0 & 0 \\ 0 & 0 & 0 & 0 & 1 & x_1 & x_1y_1 & y_1 & 0 & 0 & 0 & 0 \\ 0 & 0 & 0 & 0 & 0 & 0 & 0 & 0 & 1 & x_1 & x_1y_1 & y_1 \\ 1 & x_2 & x_2y_2 & y_2 & 0 & 0 & 0 & 0 & 0 & 0 & 0 & 0 \\ 0 & 0 & 0 & 0 & 1 & x_2 & x_2y_2 & y_2 & 0 & 0 & 0 & 0 \\ 0 & 0 & 0 & 0 & 0 & 0 & 0 & 0 & 1 & x_2 & x_2y_2 & y_2 \\ 1 & x_3 & x_3y_3 & y_3 & 0 & 0 & 0 & 0 & 0 & 0 & 0 & 0 \\ 0 & 0 & 0 & 0 & 1 & x_3 & x_3y_3 & y_3 & 0 & 0 & 0 & 0 \\ 0 & 0 & 0 & 0 & 0 & 0 & 0 & 0 & 1 & x_3 & x_3y_3 & y_3 \\ 1 & x_4 & x_4y_4 & y_4 & 0 & 0 & 0 & 0 & 0 & 0 & 0 & 0 \\ 0 & 0 & 0 & 0 & 1 & x_4 & x_4y_4 & 0 & 0 & 0 & 0 & 0 \\ 0 & 0 & 0 & 0 & 0 & 0 & 0 & y_4 & 1 & y_4 & x_4y_4 & y_4 \end{bmatrix} \begin{bmatrix} c_1 \\ c_2 \\ c_3 \\ c_4 \\ c_5 \\ c_6 \\ c_7 \\ c_8 \\ c_9 \\ c_{10} \\ c_{11} \\ c_{12} \end{bmatrix}$$

$$= \bar{A}_6 \bar{c}_6 \tag{6.40}$$

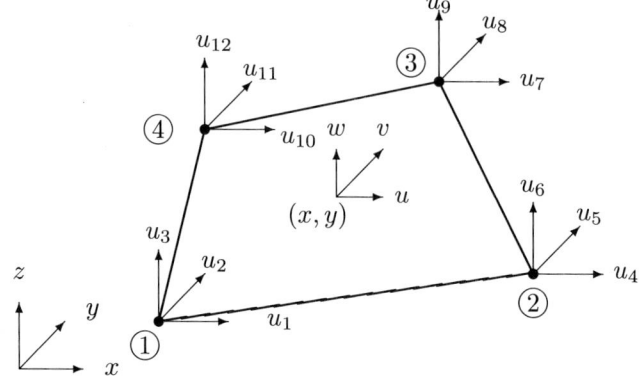

Fig. 6.3   Node displacements of a quadrilateral plate used for the mass matrix.

where $\bar{\mathbf{A}}_5$ is the (12×12) matrix in Eq. (6.38). Therefore,

$$\bar{\mathbf{c}}_6 = \bar{\mathbf{A}}_6^{-1}\bar{\mathbf{u}} \tag{6.41}$$

and

$$\begin{bmatrix} u \\ v \\ w \end{bmatrix} = \bar{\mathbf{C}}_6\bar{\mathbf{A}}_6^{-1}\bar{\mathbf{u}} = \bar{\mathbf{a}}_5\bar{\mathbf{u}} \tag{6.42}$$

where

$$\bar{\mathbf{a}}_6 = \bar{\mathbf{C}}_6\bar{\mathbf{A}}_5^{-1} \tag{6.43}$$

is the shape function and

$$\bar{\mathbf{u}} = \{u_1 \cdots u_{12}\} \tag{6.44}$$

The mass matrix $\mathbf{m}_6$ for this element is then determined from

$$\mathbf{m}_6 = \rho \iiint \bar{\mathbf{a}}_6^T\bar{\mathbf{a}}_6 \, dx \, dy \, dz = \rho t \int_{-1}^{1} \int_{-1}^{1} \bar{\mathbf{a}}_5^T\bar{\mathbf{a}}_6 |\mathbf{J}(\xi, \eta)| \, d\xi \, d\eta \tag{6.45}$$

where $|\mathbf{J}(x, y)|$ is the Jacobian introduced earlier and the integration can be performed using the Gaussian quadrature. As an approximation, the thermal stiffness, thermal load, and mass matrix can be assumed to be the same as for the T5 element, that is,

$$\mathbf{h}_6 \approx \mathbf{h}_5 \tag{6.46}$$

$$\mathbf{q}_6 \approx \mathbf{q}_5 \tag{6.47}$$

$$\mathbf{m}_6 \approx \mathbf{m}_5 \tag{6.48}$$

Part 3
Solid and Basic Elements

9

Tetrahedron T 7: Assumed Displacement Distribution

The node numbering and element displacements $u_1 \cdots u_{12}$ for a tetrahedron element are shown in Fig. 7.1, where the origin for the rectangular coordinates is assumed to be at node no. 1. At each node point there are three element forces and three displacements in the x, y, and z directions, respectively.

For each displacement u_i there is a corresponding element force S_i, where $i = 1 \cdots 12$. The displacement field for the tetrahedron element can be represented by

$$u = c_1 + c_2 x + c_3 y + c_4 z \tag{7.1}$$

$$v = c_5 + c_6 x + c_7 y + c_8 z \tag{7.2}$$

$$w = c_9 + c_{10} x + c_{11} y + c_{12} z \tag{7.3}$$

where u, v, and w are displacements at a point (x, y, z) in the x, y, and z directions, respectively (Fig. 7.1). *To simplify the analysis, the origin for the rectangular coordinates will be taken at node 1 with the y axis coinciding with the element edge from node 1 to node 3.*

Hence the element node displacements $\mathbf{u}$ can be expressed as

$$\mathbf{u} = \begin{bmatrix} u_1 \\ u_2 \\ u_3 \\ u_4 \\ u_5 \\ u_6 \\ u_7 \\ u_8 \\ u_9 \\ u_{10} \\ u_{11} \\ u_{12} \end{bmatrix} = \begin{bmatrix} 1 & x_1 & y_1 & z_1 & 0 & 0 & 0 & 0 & 0 & 0 & 0 & 0 \\ 0 & 0 & 0 & 0 & 1 & x_1 & y_1 & z_1 & 0 & 0 & 0 & 0 \\ 0 & 0 & 0 & 0 & 0 & 0 & 0 & 0 & 1 & x_1 & y_1 & z_1 \\ 1 & x_2 & y_2 & z_2 & 0 & 0 & 0 & 0 & 0 & 0 & 0 & 0 \\ 0 & 0 & 0 & 0 & 1 & x_2 & y_2 & z_2 & 0 & 0 & 0 & 0 \\ 0 & 0 & 0 & 0 & 0 & 0 & 0 & 0 & 1 & x_2 & y_2 & z_2 \\ 1 & x_3 & y_3 & z_3 & 0 & 0 & 0 & 0 & 0 & 0 & 0 & 0 \\ 0 & 0 & 0 & 0 & 1 & x_3 & y_3 & z_3 & 0 & 0 & 0 & 0 \\ 0 & 0 & 0 & 0 & 0 & 0 & 0 & 0 & 1 & x_3 & y_3 & z_3 \\ 1 & x_4 & y_4 & z_4 & 0 & 0 & 0 & 0 & 0 & 0 & 0 & 0 \\ 0 & 0 & 0 & 0 & 1 & x_4 & y_4 & z_4 & 0 & 0 & 0 & 0 \\ 0 & 0 & 0 & 0 & 0 & 0 & 0 & 0 & 1 & x_4 & y_4 & z_4 \end{bmatrix} \begin{bmatrix} c_1 \\ c_2 \\ c_3 \\ c_4 \\ c_5 \\ c_6 \\ c_7 \\ c_8 \\ c_9 \\ c_{10} \\ c_{11} \\ c_{12} \end{bmatrix} \tag{7.4}$$

Symbolically the preceding equation can be expressed as

$$\mathbf{u} = \mathbf{C}_7 \mathbf{c} \tag{7.5}$$

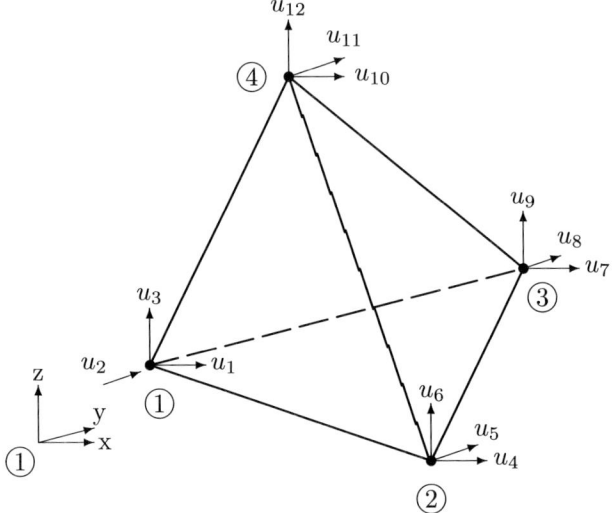

Fig. 7.1   Node displacements for tetrahedron elements.

Hence

$$\mathbf{c} = \mathbf{C}_7^{-1}\mathbf{u} \qquad (7.6)$$

The strains **e** are then computed from

$$
\mathbf{e} = \begin{bmatrix} e_{xx} \\ e_{yy} \\ e_{zz} \\ e_{xy} \\ e_{yz} \\ e_{zx} \end{bmatrix} = \begin{bmatrix} \partial u/\partial x \\ \partial v/\partial y \\ \partial w/\partial z \\ \partial v/\partial x + \partial u/\partial y \\ \partial w/\partial y + \partial v/\partial z \\ \partial u/\partial z + \partial w/\partial x \end{bmatrix}
$$

$$
= \begin{bmatrix} 0 & 1 & 0 & 0 & 0 & 0 & 0 & 0 & 0 & 0 & 0 & 0 \\ 0 & 0 & 0 & 0 & 0 & 0 & 1 & 0 & 0 & 0 & 0 & 0 \\ 0 & 0 & 0 & 0 & 0 & 0 & 0 & 0 & 0 & 0 & 0 & 1 \\ 0 & 0 & 1 & 0 & 0 & 1 & 0 & 0 & 0 & 0 & 0 & 0 \\ 0 & 0 & 0 & 0 & 0 & 0 & 0 & 1 & 0 & 0 & 1 & 0 \\ 0 & 0 & 0 & 1 & 0 & 0 & 0 & 0 & 0 & 1 & 0 & 0 \end{bmatrix} \begin{bmatrix} c_1 \\ c_2 \\ \vdots \\ c_{11} \\ c_{12} \end{bmatrix} \qquad (7.7)
$$

or symbolically as

$$\mathbf{e} = \mathbf{A}_7\mathbf{c} = \mathbf{A}_7\mathbf{C}_7^{-1}\mathbf{u} = \mathbf{b}_7\mathbf{u} \qquad (7.8)$$

where

$$\mathbf{b}_7 = \mathbf{A}_7\mathbf{C}_7^{-1} \qquad (7.9)$$

Here the matrix $\mathbf{b}_7$ represents a matrix of strains caused by unit displacements at the element nodes. These strains satisfy equations of strain compatibility and stress equilibrium within the element. All six components of strains are constant throughout the tetrahedron element. The stress-strain equation for the three-dimensional stress field can be written symbolically as

$$\sigma = \mathbf{E}_3 \mathbf{e} - \frac{\alpha T E}{(1 - 2v)} \{1\ 1\ 1\ 0\ 0\ 0\} \tag{7.10}$$

and the expression for the three-dimensional Young's modulus is given by

$$\mathbf{E}_3 = \frac{E}{(1 + v)(1 - 2v)}$$

$$\times \begin{bmatrix} (1-v) & v & v & 0 & 0 & 0 \\ v & (1-v) & v & 0 & 0 & 0 \\ v & v & (1-v) & 0 & 0 & 0 \\ 0 & 0 & 0 & (1-2v) & 0 & 0 \\ 0 & 0 & 0 & 0 & (1-2v) & 0 \\ 0 & 0 & 0 & 0 & 0 & (1-2v) \end{bmatrix}$$

$$\tag{7.11}$$

The stiffness matrix for the tetrahedron element is obtained from

$$\mathbf{k}_7 = \int_V \mathbf{b}_7^T \mathbf{E}_3 \mathbf{b}_7 \ dV$$

$$= \iiint \mathbf{b}_7^T \mathbf{E}_3 \mathbf{b}_7 \ dx\, dy\, dz = \mathbf{b}_7^T \mathbf{E}_3 \mathbf{b}_7 \int dV = \mathbf{b}_7^T \mathbf{E}_3 \mathbf{b}_7 V \tag{7.12}$$

where V is the element volume calculated from the simple formula

$$V = \frac{1}{6} \begin{vmatrix} 1 & x_1 & y_1 & z_1 \\ 1 & x_2 & y_2 & z_2 \\ 1 & x_3 & y_3 & z_3 \\ 1 & x_4 & y_4 & z_4 \end{vmatrix} = \frac{1}{6} x_2 y_2 z_4 \tag{7.13}$$

because $x_1 = y_1 = 0$, $z_1 = 0$, and $x_3 = 0$.

Here $\mathbf{b}_7$ in this case is a matrix of constant coefficients, and therefore the product $\mathbf{b}_7^T \mathbf{E}_3 \mathbf{b}_7$ can be taken outside the volume integral. As an example, the stiffness matrix $\mathbf{k}_7$ for $v = 0.3$ and $x_1 = 0$, $y_1 = 0$, $z_1 = 0$, $x_2 = 10$ in., $y_2 = 0$, $z_2 = 0$, $x_3 = 0$, $y_3 = 10$ in., $z_3 = 0$ and $x_4 = 0$, $y_4 = 0$ and $z_4 = 10$ in. is given in Eq. (7.14):

$$\mathbf{k}_7 = EV \times \begin{bmatrix}
4.8077 & 2.2436 & 2.2436 & -2.2436 & -1.2821 & -1.2821 \\
2.2436 & 4.8077 & 2.2436 & -0.9615 & -1.2821 & 0.0000 \\
2.2436 & 2.2436 & 4.8077 & -0.9615 & 0.0000 & -1.2821 \\
-2.2436 & -0.9615 & -0.9615 & 2.2436 & 0.0000 & 0.0000 \\
-1.2821 & -1.2821 & 0.0000 & 0.0000 & 1.2821 & 0.0000 \\
-1.2821 & 0.0000 & -1.2821 & 0.0000 & 0.0000 & 1.2821 \\
-1.2821 & -1.2821 & 0.0000 & 0.0000 & 1.2821 & 0.0000 \\
-0.9615 & -2.2436 & -0.9615 & 0.9615 & 0.0000 & 0.0000 \\
0.0000 & -1.2821 & -1.2821 & 0.0000 & 0.0000 & 0.0000 \\
-1.2821 & 0.0000 & -1.2821 & 0.0000 & 0.0000 & 1.2821 \\
0.0000 & -1.2821 & -1.2821 & 0.0000 & 0.0000 & 0.0000 \\
-0.9615 & -0.9615 & -2.2436 & 0.9615 & 0.0000 & 0.0000
\end{bmatrix}$$

$$\begin{bmatrix}
-1.2821 & -0.9615 & 0.0000 & -1.2821 & 0.0000 & -0.9615 \\
-1.2821 & -2.2436 & -1.2821 & 0.0000 & -1.2821 & -0.9615 \\
0.0000 & -0.9615 & -1.2821 & -1.2821 & -1.2821 & -2.2436 \\
0.0000 & 0.9615 & 0.0000 & 0.0000 & 0.0000 & 0.9615 \\
1.2821 & 0.0000 & 0.0000 & 0.0000 & 0.0000 & 0.0000 \\
0.0000 & 0.0000 & 0.0000 & 1.2821 & 0.0000 & 0.0000 \\
1.2821 & 0.0000 & 0.0000 & 0.0000 & 0.0000 & 0.0000 \\
0.0000 & 2.2436 & 0.0000 & 0.0000 & 0.0000 & 0.9615 \\
0.0000 & 0.0000 & 1.2821 & 0.0000 & 1.2821 & 0.0000 \\
0.0000 & 0.0000 & 0.0000 & 1.2821 & 0.0000 & 0.0000 \\
0.0000 & 0.0000 & 1.2821 & 0.0000 & 1.2821 & 0.0000 \\
0.0000 & 0.9615 & 0.0000 & 0.0000 & 0.0000 & 2.2436
\end{bmatrix}$$

$$(7.14)$$

Thermal Stiffness

The thermal stiffness is computed from

$$\mathbf{h}_7 = \int_v \mathbf{b}_7^T \frac{E}{(1-2v)} \begin{bmatrix} -1 \\ -1 \\ -1 \\ 0 \\ 0 \\ 0 \end{bmatrix} dV = \frac{EV\mathbf{b}_7^T}{(1-2v)} \begin{bmatrix} -1 \\ -1 \\ -1 \\ 0 \\ 0 \\ 0 \end{bmatrix} = \frac{EV}{10(1-2v)} \begin{bmatrix} 1 \\ 1 \\ 1 \\ -1 \\ 0 \\ 0 \\ 0 \\ 0 \\ -1 \\ 0 \\ 0 \\ 0 \\ -1 \end{bmatrix}$$

$$(7.15)$$

where the right-hand side of Eq. (7.15) shows the numerical values from the preceding example. The force system resulting from $\mathbf{h}_7$ is self-equilibrating.

Thermal Load

The thermal load q_7 is obtained from

$$\mathbf{q}_7 = \mathbf{h}_7 \alpha T \tag{7.16}$$

where T is the element temperature and α is the coefficient of thermal expansion.

Mass Matrix

Equations (7.1–7.3) can be combined to give

$$\begin{bmatrix} u \\ v \\ w \end{bmatrix} = \begin{bmatrix} 1 & x & y & z & 0 & 0 & 0 & 0 & 0 & 0 & 0 & 0 \\ 0 & 0 & 0 & 0 & 1 & x & y & z & 0 & 0 & 0 & 0 \\ 0 & 0 & 0 & 0 & 0 & 0 & 0 & 0 & 1 & x & y & z \end{bmatrix} \begin{bmatrix} c_1 \\ \vdots \\ c_{12} \end{bmatrix} = \mathbf{D}_7 \mathbf{c} \tag{7.17}$$

where $\mathbf{D}_7$ is the preceding (3×12) matrix. Substituting Eq. (7.6) into Eq. (7.17)

$$\begin{bmatrix} u \\ v \\ w \end{bmatrix} = \mathbf{D}_7 \mathbf{C}_7^{-1} \mathbf{u} = \mathbf{a}_7 \mathbf{u} \tag{7.18}$$

where

$$\mathbf{a}_7 = \mathbf{D}_7 \mathbf{C}_7^{-1} \tag{7.19}$$

is the shape function for the tetrahedron. The mass matrix for the tetrahedron is calculated from

$$\mathbf{m}_7 = \rho \int_v \mathbf{a}_7^T \mathbf{a}_7 \, dV = \frac{\rho V}{20} \begin{bmatrix} 2 & 0 & 0 & 1 & 0 & 0 & 1 & 0 & 0 & 1 & 0 & 0 \\ 0 & 2 & 0 & 0 & 1 & 0 & 0 & 1 & 0 & 0 & 1 & 0 \\ 0 & 0 & 2 & 0 & 0 & 1 & 0 & 0 & 1 & 0 & 0 & 1 \\ 1 & 0 & 0 & 2 & 0 & 0 & 1 & 0 & 0 & 1 & 0 & 0 \\ 0 & 1 & 0 & 0 & 2 & 0 & 0 & 1 & 0 & 0 & 1 & 0 \\ 0 & 0 & 1 & 0 & 0 & 2 & 0 & 0 & 1 & 0 & 0 & 1 \\ 1 & 0 & 0 & 1 & 0 & 0 & 2 & 0 & 0 & 1 & 0 & 0 \\ 0 & 1 & 0 & 0 & 1 & 0 & 0 & 2 & 0 & 0 & 1 & 0 \\ 0 & 0 & 1 & 0 & 0 & 1 & 0 & 0 & 2 & 0 & 0 & 1 \\ 1 & 0 & 0 & 1 & 0 & 0 & 1 & 0 & 0 & 2 & 0 & 0 \\ 0 & 1 & 0 & 0 & 1 & 0 & 0 & 1 & 0 & 0 & 2 & 0 \\ 0 & 0 & 1 & 0 & 0 & 1 & 0 & 0 & 1 & 0 & 0 & 2 \end{bmatrix} \tag{7.20}$$

10

Tetrahedron T8: Assumed Displacement Distribution plus Corrective Distribution Inside the Element Boundaries

To obtain greater accuracy beyond that of the constant strain for a tetrahedron element, additional corrective displacements that vanish on the boundaries are introduced. The node numbering and element displacements $u_1 \cdots u_{12}$ for a tetrahedron element are shown in Fig. 8.1, where the origin for the rectangular coordinates is assumed to be at node no. 1. At each node point there are three element forces and three displacements in the x, y, and z directions, respectively. For each displacement u_i there is a corresponding element force S_i, where $i = 1 \cdots 12$. The displacement field for the tetrahedron element can be represented by

$$u = c_1 + c_2 x + c_3 y + c_4 z \tag{8.1}$$

$$v = c_5 + c_6 x + c_7 y + c_8 z \tag{8.2}$$

$$w = c_9 + c_{10} x + c_{11} y + c_{12} z \tag{8.3}$$

where u, v, and w are displacements at a point (x, y, z) in the x, y, and z directions, respectively. *To simplify the analysis, the origin for the rectangular coordinates will be taken at the node 1 with the y axis coinciding with the element edge from node 1 to node 3 and the x axis in the plane of the tetrahedron base.* The displacements will be evaluated in terms of the tetrahedral coordinates shown in Figs. 8.2–8.4. The x, y, and z coordinates are functions of the ξ, η, and ζ tetrahedral coordinates as shown in Eqs. (8.5–8.7). The basic displacement distribution represented by Eqs. (8.1–8.3) can be augmented by a correction function vanishing on the boundaries of the tetrahedron element. A simple function f in terms of the tetrahedral coordinates ξ, η, and ζ can be taken as

$$f = f(\xi, \eta, \zeta) = \xi(1 - \xi)\eta(1 - \eta)\zeta(1 - \zeta)$$

$$= (\xi - \xi^2)(\eta - \eta^2)(\zeta - \zeta^2) \tag{8.4}$$

The relationship between the rectangular and a special tetrahedral coordinate sytems is given by

$$x = \xi(1 - \eta)x_2\zeta + (1 - \zeta)x_4 \tag{8.5}$$

$$y = \xi(y_2 - \eta y_{23})\zeta + (1 - \zeta)y_4 \tag{8.6}$$

$$z = (1 - \zeta)z_4 \tag{8.7}$$

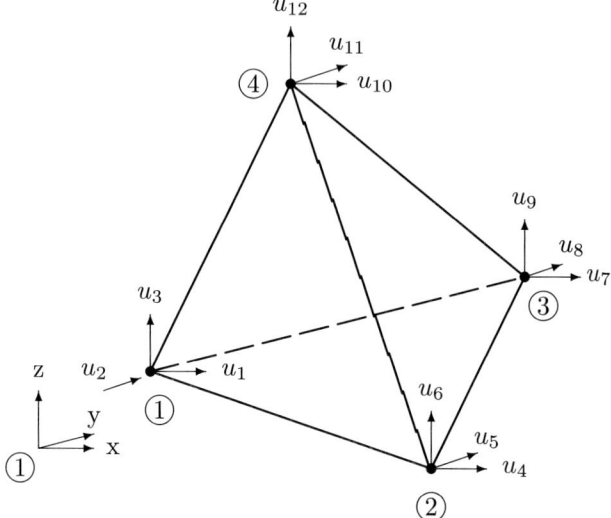

Fig. 8.1   Node displacements for tetrahedron elements.

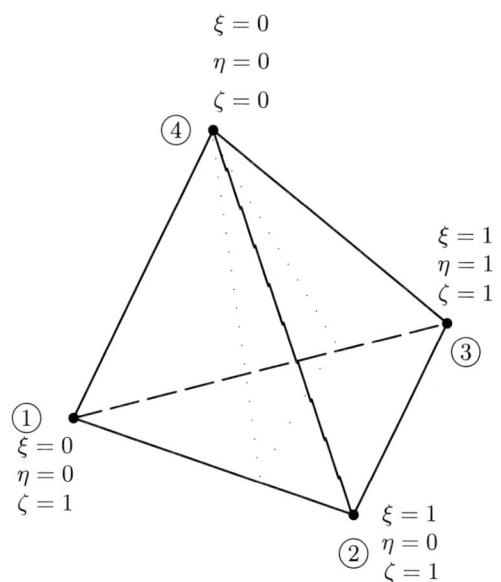

Fig. 8.2   Tetrahedral coordinate system: $\xi = $ constant.

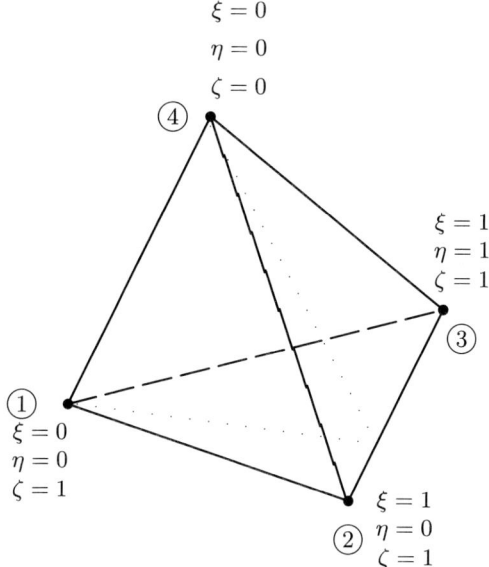

Fig. 8.3   Tetrahedral coordinate system: $\eta = $ **constant.**

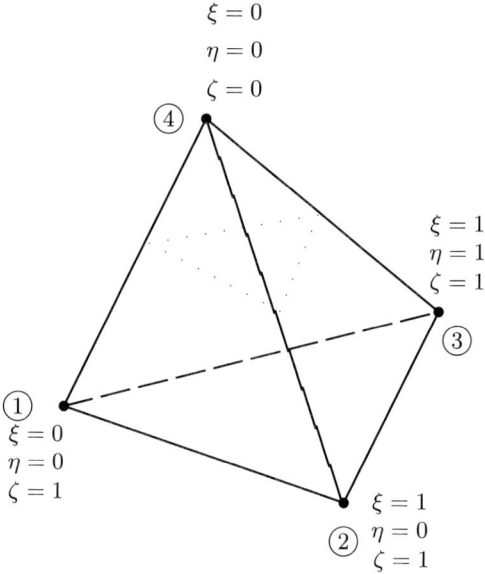

Fig. 8.4   Tetrahedral coordinate system: $\zeta = $ **constant.**

where

$$y_{23} = (y_2 - y_3) \tag{8.8}$$

The location of the four node points is given in rectangular coordinates as $(x_1, y_1, z_1), (x_2, y_2, z_2), (x_3, y_3, z_3), (x_4, y_4, z_4)$, and the preceding equations can now be used to determine expressions for ξ, η, and ζ as

$$\xi = \frac{x - (1 - \zeta)x_4}{(1 - \eta)x_2\zeta} \tag{8.9}$$

$$\eta = \frac{(\xi y_2 - y)\zeta}{\xi y_{23}} \tag{8.10}$$

$$\zeta = \frac{1 - z}{z_4} \tag{8.11}$$

The function $f(\xi, \eta, \zeta)$ vanishes on the boundaries of the tetrahedron element. Thus the new displacement field for the tetrahedron becomes

$$u = c_1 + c_2 x + c_3 y + c_4 z + c_{13} f(\xi, \eta, \zeta) \tag{8.12}$$

$$v = c_5 + c_6 x + c_7 y + c_8 z + c_{14} f(\xi, \eta, \zeta) \tag{8.13}$$

$$w = c_9 + c_{10} x + c_{11} y + c_{12} z + c_{15} f(\xi, \eta, \zeta) \tag{8.14}$$

To obtain the element strains, the partial derivatives of the displacements must be obtained first. To find the partial derivatives of the function $f(\xi, \eta, \zeta)$, the chain rule of differentiation shown next is used:

$$\frac{\partial}{\partial x} = \frac{\partial}{\partial \xi}\frac{\partial \xi}{\partial x} + \frac{\partial}{\partial \eta}\frac{\partial \eta}{\partial x} + \frac{\partial}{\partial \zeta}\frac{\partial \zeta}{\partial x} \tag{8.15}$$

$$\frac{\partial}{\partial y} = \frac{\partial}{\partial \xi}\frac{\partial \xi}{\partial y} + \frac{\partial}{\partial \eta}\frac{\partial \eta}{\partial y} + \frac{\partial}{\partial \zeta}\frac{\partial \zeta}{\partial y} \tag{8.16}$$

$$\frac{\partial}{\partial z} = \frac{\partial}{\partial \xi}\frac{\partial \xi}{\partial z} + \frac{\partial}{\partial \eta}\frac{\partial \eta}{\partial z} + \frac{\partial}{\partial \zeta}\frac{\partial \zeta}{\partial z} \tag{8.17}$$

From the relationship between the tetrahedral and rectangular cordinates in Eqs. (8.9–8.11), it follows that

$$\frac{\partial \xi}{\partial x} = \frac{1}{x_2(1 - \eta)\zeta} \tag{8.18}$$

$$\frac{\partial \eta}{\partial x} = 0 \tag{8.19}$$

$$\frac{\partial \zeta}{\partial x} = 0 \tag{8.20}$$

$$\frac{\partial \xi}{\partial y} = 0 \tag{8.21}$$

$$\frac{\partial \eta}{\partial y} = -\frac{\zeta}{\xi y_{23}} \tag{8.22}$$

$$\frac{\partial \zeta}{\partial y} = 0 \tag{8.23}$$

$$\frac{\partial \xi}{\partial z} = 0 \tag{8.24}$$

$$\frac{\partial \eta}{\partial z} = 0 \tag{8.25}$$

$$\frac{\partial \zeta}{\partial z} = -\frac{1}{z_4} \tag{8.26}$$

Also from Eq. (8.4) the following relations are obtained:

$$\frac{\partial f}{\partial \xi} = (1 - 2\xi)\eta(1 - \eta)\zeta(1 - \zeta) \tag{8.27}$$

$$\frac{\partial f}{\partial \eta} = \xi(1 - \xi)(1 - 2\eta)\zeta(1 - \zeta) \tag{8.28}$$

$$\frac{\partial f}{\partial \zeta} = \xi(1 - \xi)\eta(1 - \eta)(1 - 2\zeta) \tag{8.29}$$

Using now the relations in Eqs. (8.18–8.29) and (8.12–8.14)

$$\frac{\partial f}{\partial x} = \frac{\partial f}{\partial \xi}\frac{\partial \xi}{\partial x} = \frac{(1 - 2\xi)\eta(1 - \eta)\zeta(1 - \zeta)}{x_2} \tag{8.30}$$

$$\frac{\partial f}{\partial y} = \frac{\partial f}{\partial \eta}\frac{\partial \eta}{\partial y} = \frac{-(1 - \xi)(1 - 2\eta)\zeta^2(1 - \zeta)}{y_{23}} \tag{8.31}$$

$$\frac{\partial f}{\partial z} = \frac{\partial f}{\partial \zeta}\frac{\partial \zeta}{\partial z} = \frac{-\xi(1 - \xi)\eta(1 - \eta)(1 - 2\zeta)}{z_4} \tag{8.32}$$

$$\begin{bmatrix} u \\ v \\ w \end{bmatrix} = \begin{bmatrix} \mathbf{G}_a & \mathbf{G}_b \end{bmatrix}\mathbf{c} = \mathbf{Gc} \tag{8.33}$$

where $\mathbf{G}$ is a rectangular matrix whose elements are the assumed displacement functions. The matrix $\mathbf{G}$ consists of two submatrices $\mathbf{G}_a$ and $\mathbf{G}_b$ given by

$$\mathbf{G}_a = \begin{bmatrix} 1 & x & y & z & 0 & 0 & 0 & 0 & 0 & 0 & 0 & 0 \\ 0 & 0 & 0 & 0 & 1 & x & y & z & 0 & 0 & 0 & 0 \\ 0 & 0 & 0 & 0 & 0 & 0 & 0 & 0 & 1 & x & y & z \end{bmatrix} \tag{8.34}$$

and

$$\mathbf{G}_b = \begin{bmatrix} f & 0 & 0 \\ 0 & f & 0 \\ 0 & 0 & f \end{bmatrix} \tag{8.35}$$

while the column matrix $\mathbf{c}$ is given by

$$\mathbf{c} = \{c_1 \; c_2 \; \cdots \; c_{15}\} = \{\mathbf{c}_a \; \mathbf{c}_b\} \tag{8.36}$$

where

$$\mathbf{c}_a = \{c_1 \cdots c_{12}\} \tag{8.37}$$

$$\mathbf{c}_b = \{c_{13} \; c_{14} \; c_{15}\} \tag{8.38}$$

Using again Eqs. (8.12–8.14) and noting that $z_1 = 0, z_2 = 0,$ and $z_3 = 0,$ the node displacements $\mathbf{u}$ can be expressed as

$$\mathbf{u} = \begin{bmatrix} u_1 \\ u_2 \\ u_3 \\ u_4 \\ u_5 \\ u_6 \\ u_7 \\ u_8 \\ u_9 \\ u_{10} \\ u_{11} \\ u_{12} \end{bmatrix} = \begin{bmatrix} 1 & 0 & 0 & 0 & 0 & 0 & 0 & 0 & 0 & 0 & 0 & 0 \\ 0 & 0 & 0 & 0 & 1 & 0 & 0 & 0 & 0 & 0 & 0 & 0 \\ 0 & 0 & 0 & 0 & 0 & 0 & 0 & 0 & 1 & 0 & 0 & 0 \\ 1 & x_2 & y_2 & 0 & 0 & 0 & 0 & 0 & 0 & 0 & 0 & 0 \\ 0 & 0 & 0 & 0 & 1 & x_2 & y_2 & 0 & 0 & 0 & 0 & 0 \\ 0 & 0 & 0 & 0 & 0 & 0 & 0 & 0 & 1 & x_2 & 0 & 0 \\ 1 & x_3 & y_3 & 0 & 0 & 0 & 0 & 0 & 0 & 0 & 0 & 0 \\ 0 & 0 & 0 & 0 & 1 & x_3 & y_3 & 0 & 0 & 0 & 0 & 0 \\ 0 & 0 & 0 & 0 & 0 & 0 & 0 & 0 & 1 & x_3 & y_3 & 0 \\ 1 & x_4 & y_4 & z_4 & 0 & 0 & 0 & 0 & 0 & 0 & 0 & 0 \\ 0 & 0 & 0 & 0 & 1 & x_4 & y_4 & z_4 & 0 & 0 & 0 & 0 \\ 0 & 0 & 0 & 0 & 0 & 0 & 0 & 0 & 1 & x_4 & y_4 & z_4 \end{bmatrix} \begin{bmatrix} c_1 \\ c_2 \\ c_3 \\ c_4 \\ c_5 \\ c_6 \\ c_7 \\ c_8 \\ c_9 \\ c_{10} \\ c_{11} \\ c_{12} \end{bmatrix} \tag{8.39}$$

Symbolically, Eq. (8.39) can be expressed as

$$\mathbf{u} = \mathbf{C}_a \mathbf{c}_a \tag{8.40}$$

Hence

$$\mathbf{c}_a = \mathbf{C}_a^{-1} \mathbf{u} \tag{8.41}$$

The strains in the element can now be computed from

$$e_{xx} = \frac{\partial u}{\partial x} = \frac{c_2 + c_{13}(1 - 2\xi)\eta(1 - \eta)\zeta(1 - \zeta)}{x_2} \tag{8.42}$$

$$e_{yy} = \frac{\partial v}{\partial y} = \frac{c_7 - c_{14}\xi(1 - \xi)(1 - 2\eta)\zeta^2(1 - \zeta)}{y_{23}} \tag{8.43}$$

$$e_{zz} = \frac{\partial w}{\partial z} = \frac{c_{12} - c_{15}\xi(1 - \xi)\eta(1 - \eta)(1 - 2\zeta)}{z_4} \tag{8.44}$$

$$e_{xy} = \frac{\partial v}{\partial x} + \frac{\partial u}{\partial y} = \frac{c_3 + c_6 + c_{13}(1 - 2\xi)\eta(1 - \eta)\zeta(1 - \zeta)}{(x_2 - c_{14}\xi(1 - \xi)(1 - 2\eta)\zeta^2(1 - \zeta))/y_{23}} \tag{8.45}$$

$$e_{yz} = \frac{\partial w}{\partial y} + \frac{\partial v}{\partial z} = \frac{c_8 + c_{11} - c_{14}\xi(1-\xi)(1-2\eta)(1-2\zeta)}{y_{23} - c_{15}\xi(1-\xi)\eta(1-\eta)(1-2\zeta)/z_4} \tag{8.46}$$

$$e_{zx} = \frac{\partial u}{\partial z} + \frac{\partial w}{\partial x} = \frac{c_4 + c_{10} - c_{13}(1-2\xi)\eta(1-\eta)\zeta(-\zeta)}{(x_2 - c_{15}\xi(1-\xi)\eta)(1-\eta(1-2\zeta))/z_4} \tag{8.47}$$

which leads to

$$\mathbf{e} = \begin{bmatrix} e_{xx} \\ e_{yy} \\ e_{zz} \\ e_{xy} \\ e_{yz} \\ e_{zx} \end{bmatrix} = \mathbf{Hc} = \begin{bmatrix} \mathbf{H}_a & \mathbf{H}_b \end{bmatrix} \begin{bmatrix} \mathbf{c}_a \\ \mathbf{c}_b \end{bmatrix} \tag{8.48}$$

where

$$\mathbf{H} = \begin{bmatrix} \mathbf{H}_a & \mathbf{H}_b \end{bmatrix} \tag{8.49}$$

$$\mathbf{H}_a = \begin{bmatrix} 0 & 1 & 0 & 0 & 0 & 0 & 0 & 0 & 0 & 0 & 0 & 0 \\ 0 & 0 & 0 & 0 & 0 & 0 & 1 & 0 & 0 & 0 & 0 & 0 \\ 0 & 0 & 0 & 0 & 0 & 0 & 0 & 0 & 0 & 0 & 0 & 1 \\ 0 & 0 & 1 & 0 & 0 & 1 & 0 & 0 & 0 & 0 & 0 & 0 \\ 0 & 0 & 0 & 0 & 0 & 0 & 0 & 1 & 0 & 0 & 1 & 0 \\ 0 & 0 & 0 & 1 & 0 & 0 & 0 & 0 & 0 & 1 & 0 & 0 \end{bmatrix} \tag{8.50}$$

$$\mathbf{H}_b = \begin{bmatrix} (1-2\xi)\eta(1-\eta)\zeta(1-\zeta)x_2 & 0 \\ 0 & -\xi(1-\xi)(1-2\eta)\zeta^2(1-\zeta)/y_{23} \\ 0 & 0 \\ (1-2\xi)\eta(1-\eta)\zeta(1-\zeta)/x_2 & -\xi(1-\xi)(1-2\eta)\zeta^2(1-\zeta)/y_{23} \\ 0 & -\xi(1-\xi)(1-2\eta)(1-2\zeta/y_{23} \\ -(1-2\xi)\eta(1-\eta)\zeta(1-\zeta)/x_2 & 0 \end{bmatrix}$$

$$\begin{matrix} 0 \\ 0 \\ -\xi(1-\xi)\eta(1-\eta)(1-2\zeta)/z_4 \\ 0 \\ -\xi(1-\xi)\eta(1-\eta)(1-2\eta)(1-2\zeta)/z_4 \\ -\xi(1-\xi)\eta(1-\eta)(1-2\zeta)/z_4 \end{matrix} \tag{8.51}$$

Combining now Eq. (8.41) with the identity $\mathbf{c}_b = \mathbf{c}_b$ yields

$$\mathbf{c} = \begin{bmatrix} \mathbf{c}_a \\ \mathbf{c}_b \end{bmatrix} = \begin{bmatrix} \mathbf{C}_a^{-1} & \mathbf{0} \\ \mathbf{0} & \mathbf{I} \end{bmatrix} \begin{bmatrix} \mathbf{u} \\ \mathbf{c}_b \end{bmatrix} = \mathbf{W}\hat{\mathbf{u}} \tag{8.52}$$

where

$$\mathbf{W} = \begin{bmatrix} \mathbf{C}_a^{-1} & \mathbf{0} \\ \mathbf{0} & \mathbf{I} \end{bmatrix} \tag{8.53}$$

and

$$\hat{\mathbf{u}} = \begin{bmatrix} \mathbf{u} \\ \mathbf{c}_b \end{bmatrix} \tag{8.54}$$

The strain energy U_i in the element is given by

$$U_i = \frac{1}{2} \int_v \mathbf{e}^T \mathbf{E}_3 \mathbf{e} \, dV = \frac{1}{2} \mathbf{c}^T \int_v \mathbf{H}^T \mathbf{E}_3 \mathbf{H} \, dV \mathbf{c} = \frac{1}{2} \hat{\mathbf{u}}^T \mathbf{W}^T \int_v \mathbf{H}^T \mathbf{E}_3 \mathbf{H} \, dV \, \mathbf{W} \hat{\mathbf{u}}$$

$$= \frac{1}{2} \hat{\mathbf{u}}^T \hat{\mathbf{k}} \hat{\mathbf{u}} \tag{8.55}$$

where

$$\hat{\mathbf{k}} = \mathbf{W}^T \int_v \mathbf{H}^T \mathbf{E}_3 \mathbf{H} \mathbf{W} \, dV = \begin{bmatrix} \mathbf{k}_{aa} & \mathbf{k}_{ab} \\ \mathbf{k}_{ba} & \mathbf{k}_{bb} \end{bmatrix} \tag{8.56}$$

and the expressions for the three-dimensional Young's modulus are given by

$$\mathbf{E}_3 = \frac{E}{(1+v)(1-2v)}$$

$$\times \begin{bmatrix} (1-v) & v & v & 0 & 0 & 0 \\ v & (1-v) & v & 0 & 0 & 0 \\ v & v & (1-v) & 0 & 0 & 0 \\ 0 & 0 & 0 & (1-2v) & 0 & 0 \\ 0 & 0 & 0 & 0 & (1-2v) & 0 \\ 0 & 0 & 0 & 0 & 0 & (1-2v) \end{bmatrix}$$

$$\tag{8.57}$$

The volume integral in Eq. (8.56) is calculated from

$$\int_v \mathbf{H}^T \mathbf{E}_3 \mathbf{H} \, dV = \iiint \mathbf{H}^T \mathbf{E}_3 \mathbf{H} \, dx \, dy \, dz$$

$$= \int_{\zeta=0}^{\zeta=1} \int_{\eta=0}^{\eta=1} \int_{\xi=0}^{\xi=1} \mathbf{H}^T \mathbf{E}_3 \mathbf{H}^T \, |J(x,y,z)| \, d\xi \, d\eta \, d\zeta \tag{8.58}$$

where $|J(x,y,z)|$ is the determinant of the Jacobian J given by

$$|J(x,y,z)| = \begin{vmatrix} \dfrac{\partial x}{\partial \xi} & \dfrac{\partial x}{\partial \eta} & \dfrac{\partial x}{\partial \zeta} \\ \dfrac{\partial y}{\partial \xi} & \dfrac{\partial y}{\partial \eta} & \dfrac{\partial y}{\partial \zeta} \\ \dfrac{\partial z}{\partial \xi} & \dfrac{\partial z}{\partial \eta} & \dfrac{\partial z}{\partial \zeta} \end{vmatrix} = \begin{vmatrix} (1-\eta)x_2\zeta & -\xi x_2 & \xi(1-\eta)x_2 \\ (y_2-\eta)y_{23}\zeta & -\xi y_{23}\zeta & -\eta y_{23} \\ 0 & 0 & -z_4 \end{vmatrix}$$

$$= \xi \eta^2 x_2 y_3 z_4 = 6\xi \eta^2 V_8 \tag{8.59}$$

where V_8 is the volume of the tetrahedron element. The total potential energy U in the element can be written as

$$U = U_i - \mathbf{u}^T \mathbf{S} = \frac{1}{2} \hat{\mathbf{u}}^T \hat{\mathbf{k}} \hat{\mathbf{u}} - \mathbf{u}^T \mathbf{S} = \frac{1}{2} \begin{bmatrix} \mathbf{u}^T & \mathbf{c}_b^T \end{bmatrix} \begin{bmatrix} \mathbf{k}_{aa} & \mathbf{k}_{ab} \\ \mathbf{k}_{ba} & \mathbf{k}_{bb} \end{bmatrix} \begin{bmatrix} \mathbf{u} \\ \mathbf{c}_b \end{bmatrix} \qquad (8.60)$$

where U_i is the strain energy in the element, $\mathbf{S}$ is the matrix of the element forces corresponding with the displacements $\mathbf{u}$, and the term $\mathbf{u}^T \mathbf{S}$ is the potential of the external forces. Now the condition of the minimum potential energy requires that

$$\frac{\partial U}{\partial \hat{\mathbf{u}}} = \mathbf{0} \qquad (8.61)$$

leading to

$$\frac{1}{2} \begin{bmatrix} \mathbf{k}_{aa} & \mathbf{k}_{ab} \\ \mathbf{k}_{ba} & \mathbf{k}_{bb} \end{bmatrix} \begin{bmatrix} \mathbf{u} \\ \mathbf{c}_b \end{bmatrix} - \begin{bmatrix} \mathbf{S} \\ \mathbf{0} \end{bmatrix} = \begin{bmatrix} \mathbf{0} \\ \mathbf{0} \end{bmatrix} \qquad (8.62)$$

The preceding expression can be derived by multiplying first the individual matrices in U and then performing partial differentiation with respect to $u_1, u_2, \ldots$ and then condensing the resulting equations into the matrix equation (8.62).

The component submatrices in $\hat{\mathbf{k}}$ can be obtained most conveniently from the following formulas:

$$\mathbf{A} = \begin{bmatrix} \mathbf{I}_{aa} & \mathbf{0}_{ab} \end{bmatrix} \hat{\mathbf{k}} = \begin{bmatrix} \mathbf{k}_{aa} & \mathbf{k}_{ab} \end{bmatrix} \qquad (8.63)$$

$$\mathbf{k}_{aa} = \mathbf{A} \begin{bmatrix} \mathbf{I}_{aa} \\ \mathbf{0}_{ba} \end{bmatrix} \qquad (8.64)$$

$$\mathbf{k}_{ab} = \mathbf{A} \begin{bmatrix} \mathbf{0}_{ab} \\ \mathbf{I}_{bb} \end{bmatrix} \qquad (8.65)$$

$$\mathbf{B} = \begin{bmatrix} \mathbf{0}_{ba} & \mathbf{I}_{bb} \end{bmatrix} \hat{\mathbf{k}} = \begin{bmatrix} \mathbf{k}_{ba} & \mathbf{k}_{bb} \end{bmatrix} \qquad (8.66)$$

$$\mathbf{k}_{ba} = \mathbf{B} \begin{bmatrix} \mathbf{I}_{aa} \\ \mathbf{0}_{ba} \end{bmatrix} \qquad (8.67)$$

$$\mathbf{k}_{bb} = \mathbf{B} \begin{bmatrix} \mathbf{0}_{ab} \\ \mathbf{I}_{bb} \end{bmatrix} \qquad (8.68)$$

where $\mathbf{I}$ is an identity matrix, $\mathbf{0}$ is a null matrix, $a = 12$ and $b = 3$, while a and b indicate the number of rows and columns, respectively.

The strain energy U_i in the element is given by

$$U_i = \frac{1}{2} \int_v \mathbf{e}^T \mathbf{E}_3 \mathbf{e}^T \, dV = \frac{1}{2} \mathbf{c}^T \int_v \mathbf{H}^T \mathbf{E}_3 \mathbf{H} \, dV \mathbf{c} = \frac{1}{2} \hat{\mathbf{u}}^T \mathbf{W}^T \int_v \mathbf{H}^T \mathbf{E}_3 \mathbf{H} \, dV \mathbf{W} \hat{\mathbf{u}}$$

$$= \frac{1}{2} \hat{\mathbf{u}}^T \hat{\mathbf{k}} \hat{\mathbf{u}} \qquad (8.69)$$

where

$$\hat{\mathbf{k}} = \mathbf{W}^T \int_v \mathbf{H}^T \mathbf{E}_3 \mathbf{H} \, dV \, \mathbf{W} = \begin{bmatrix} \mathbf{k}_{aa} & \mathbf{k}_{ab} \\ \mathbf{k}_{ba} & \mathbf{k}_{bb} \end{bmatrix} \quad (8.70)$$

The matrix $\mathbf{c}_b$ can then be calculated from the second row in the preceding equation as

$$\mathbf{c}_b = -\mathbf{k}_{bb}^{-1} \mathbf{k}_{ba} \mathbf{u} \quad (8.71)$$

which when substituted into the first row in Eq. (8.62) results in

$$(\mathbf{k}_{aa} - \mathbf{k}_{ab} \mathbf{k}_{bb}^{-1} \mathbf{k}_{ba}) \mathbf{u} = \mathbf{S} \quad (8.72)$$

Hence, by definition, the element stiffness matrix $\mathbf{k}_8$ is given by

$$\mathbf{k}_8 = \mathbf{k}_{aa} - \mathbf{k}_{ab} \mathbf{k}_{bb}^{-1} \mathbf{k}_{ba} \quad (8.73)$$

As an example, the stiffness matrix $\mathbf{k}_8$ for $v = 0.3$ and $x_1 = 0$, $y_1 = 0$, $z_1 = 0$, $x_2 = 10$ in., $y_2 = 0$, $z_2 = 0$, $x_3 = 0$, $y_3 = 10$ in., z_3 and $x_4 = 0$, $y_4 = 0$ and $z_4 = 10$ in. is given here:

$$\mathbf{k}_8 = EV \times$$

$$\begin{bmatrix}
4.4707 & 1.8795 & 2.1008 & -1.9336 & -1.1273 & -1.4097 \\
1.8795 & 4.4033 & 2.0786 & -0.6379 & -1.1148 & -0.1269 \\
2.1008 & 2.0786 & 4.4447 & -0.8408 & -0.0018 & -1.2582 \\
-1.9336 & -0.6379 & -0.8408 & -1.9473 & -0.1422 & 0.1285 \\
-1.1273 & -1.1148 & -0.0018 & -0.1422 & 1.1954 & 0.0741 \\
-1.4097 & -0.1269 & -1.2582 & 0.1285 & 0.0741 & 1.2072 \\
-1.1273 & -1.1148 & -0.0018 & -0.1422 & 1.1954 & 0.0741 \\
-0.7522 & -2.0064 & -0.7948 & 0.7801 & -0.0807 & 0.0528 \\
0.0000 & -1.1282 & -1.2821 & 0.0000 & 0.0000 & 0.0000 \\
-1.4097 & -1.1269 & -1.2582 & 0.1285 & 0.0741 & 1.2072 \\
0.0000 & -1.2821 & -1.2821 & 0.0000 & 0.0000 & 0.0000 \\
-0.6910 & -0.6697 & -1.9045 & 0.7123 & -0.0723 & 0.0510
\end{bmatrix}$$

$$\begin{bmatrix}
-1.1273 & -0.7522 & 0.0000 & -1.4097 & 0.0000 & -0.6910 \\
-1.1148 & -2.0064 & -1.2821 & -0.1269 & -1.2821 & -0.6697 \\
-0.0018 & -0.7948 & -1.2821 & -1.2582 & -1.2821 & -1.9045 \\
-0.1422 & 0.7801 & 0.0000 & 0.1285 & 0.0000 & 0.7123 \\
1.1954 & -0.0807 & 0.0000 & 0.0741 & 0.0000 & -0.0723 \\
0.0741 & 0.0528 & 0.0000 & 1.2072 & 0.0000 & 0.0510 \\
1.1954 & -0.0807 & 0.0000 & 0.0741 & 0.0000 & -0.0723 \\
-0.0807 & 2.0871 & 0.0000 & 0.0528 & 0.0000 & 0.7420 \\
0.0000 & 0.0000 & 1.2821 & 0.0000 & 1.2821 & 0.0000 \\
0.0741 & 0.0528 & 0.0000 & 1.2072 & 0.0000 & 0.0510 \\
0.0000 & 0.0000 & 1.2821 & 0.0000 & 1.2821 & 0.0000 \\
-0.0723 & 0.7420 & 0.0000 & 0.0510 & 0.0000 & 1.8534
\end{bmatrix}$$

$$(8.74)$$

The stress in the element can now be calculated from

$$\sigma = \mathbf{E}_3 \mathbf{e} \quad (8.75)$$

Thermal Stiffness, Thermal Loads, and Mass Matrix

As an approximation, the thermal stiffness, thermal load, and mass matrix can be assumed to be the same as for the T7 element, that is,

$$\mathbf{h}_8 \approx \mathbf{h}_7 \tag{8.76}$$

$$\mathbf{q}_8 \approx \mathbf{q}_7 \tag{8.77}$$

$$\mathbf{m}_8 \approx \mathbf{m}_7 \tag{8.78}$$

11

Prismatic Pentahedron T9: Assumed Displacement Distribution

The element displacements for a pentahedron element are shown in Fig. 9.1. The element forces (not shown in this figure) are in the same directions as the corresponding element displacements. The displacements within the element u, v, and w in the x, y, and z directions are assumed to be given by

$$u = c_1 + c_2 x + c_3 y + c_4 z + c_5 xy + c_6 zx \qquad (9.1)$$

$$v = c_7 + c_8 x + c_9 y + c_{10} z + c_{11} yz + c_{12} xy \qquad (9.2)$$

$$w = c_{13} + c_{14} x + c_{15} y + c_{16} z + c_{17} zx + c_{18} yz \qquad (9.3)$$

These assumed displacements ensure that all three direct strains e_{xx}, e_{yy}, and e_{zz} vary linearly in cross-wise directions. To simplify the expressions for node displacements, it can be assumed without loss of generality that the origin for the rectangular coordinate system is at node 1 and *the y axis coincides with the edge (1)–(2)*. Therefore,

$$x_1 = x_3 = x_4 = x_6 = y_1 = y_4 = z_1 = z_2 = z_3 = 0 \qquad (9.4)$$

Because the shaded triangular planes shown in Fig. 9.1 are parallel to each other and perpendicular to the z axis,

$$x_5 = x_2 \qquad (9.5)$$

$$y_5 = y_2 \qquad (9.6)$$

$$y_6 = y_3 \qquad (9.7)$$

$$z_4 = z_5 = z_6 = c \qquad (9.8)$$

where c is the length of the prismatic pentahedron. Thus using Eq. (9.1–9.3) together with the coordinate relations Eqs. (9.4–9.8), the element node

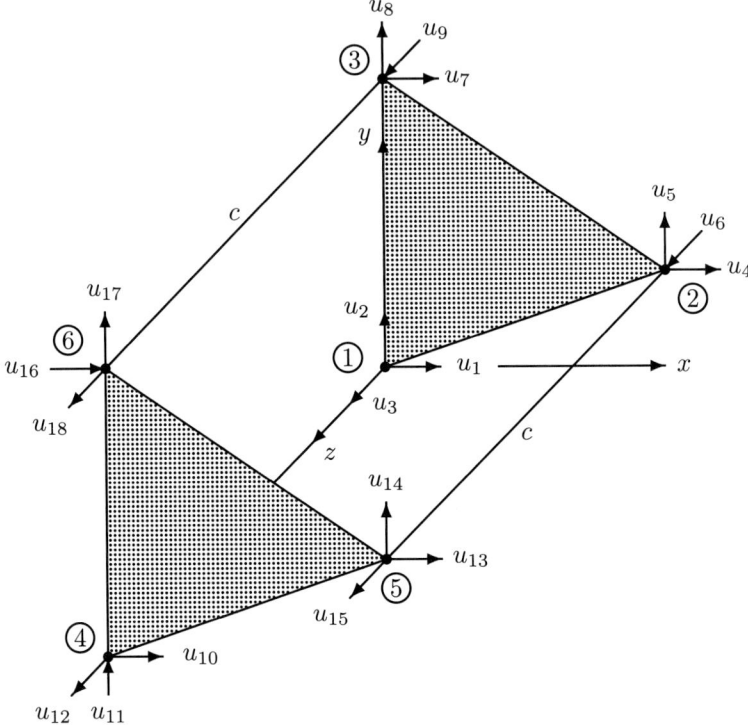

Fig. 9.1   Node displacements for prismatic pentahedrons. (The shaded sides are in the *x–y* planes.)

displacements can be expressed as

$$
\mathbf{u} =
\begin{bmatrix}
u_1 \\ u_2 \\ u_3 \\ u_4 \\ u_5 \\ u_6 \\ u_7 \\ u_8 \\ u_9 \\ u_{10} \\ u_{11} \\ u_{12} \\ u_{13} \\ u_{14} \\ u_{15} \\ u_{16} \\ u_{17} \\ u_{18}
\end{bmatrix}
=
\begin{bmatrix}
1 & 0 & 0 & 0 & 0 & 0 & 0 & 0 & 0 \\
0 & 0 & 0 & 0 & 0 & 0 & 1 & 0 & 0 \\
0 & 0 & 0 & 0 & 0 & 0 & 0 & 0 & 0 \\
1 & x_2 & y_2 & 0 & 0 & 0 & 0 & 0 & 0 \\
0 & 0 & 0 & 0 & 0 & 0 & 1 & x_2 & y_2 \\
0 & 0 & 0 & 0 & 0 & 0 & 0 & 0 & 0 \\
1 & 0 & y_3 & 0 & 0 & 0 & 0 & 0 & 0 \\
0 & 0 & 0 & 0 & 0 & 0 & 1 & 0 & y_3 \\
0 & 0 & 0 & 0 & 0 & 0 & 0 & 0 & 0 \\
1 & 0 & 0 & c & 0 & 0 & 0 & 0 & 0 \\
0 & 0 & 0 & 0 & 0 & 0 & 1 & 0 & 0 \\
0 & 0 & 0 & 0 & 0 & 0 & 0 & 0 & 0 \\
1 & x_2 & y_2 & c & y_3c & cx_2 & 0 & 0 & 0 \\
0 & 0 & 0 & 0 & 0 & 0 & 1 & x_2 & y_2 \\
0 & 0 & 0 & 0 & 0 & 0 & 0 & 0 & 0 \\
1 & 0 & y_3 & c & y_3c & 0 & 0 & 0 & 0 \\
0 & 0 & 0 & 0 & 0 & 0 & 1 & 0 & y_3 \\
0 & 0 & 0 & 0 & 0 & 0 & 0 & 0 & 0
\end{bmatrix}
$$

$$\begin{bmatrix}
0 & 0 & 0 & 0 & 0 & 0 & 0 & 0 & 0 \\
0 & 0 & 0 & 0 & 0 & 0 & 0 & 0 & 0 \\
0 & 0 & 0 & 1 & 0 & 0 & 0 & 0 & 0 \\
0 & 0 & 0 & 0 & 0 & 0 & 0 & 0 & 0 \\
0 & 0 & 0 & 0 & 0 & 0 & 0 & 0 & 0 \\
0 & 0 & 0 & 1 & x_2 & y_2 & 0 & 0 & 0 \\
0 & 0 & 0 & 0 & 0 & 0 & 0 & 0 & 0 \\
0 & 0 & 0 & 0 & 0 & 0 & 0 & 0 & 0 \\
0 & 0 & 0 & 1 & 0 & y_3 & 0 & 0 & 0 \\
0 & 0 & 0 & 0 & 0 & 0 & 0 & 0 & 0 \\
c & 0 & 0 & 0 & 0 & 0 & 0 & 0 & 0 \\
0 & 0 & 0 & 1 & 0 & 0 & c & 0 & 0 \\
0 & 0 & 0 & 0 & 0 & 0 & 0 & 0 & 0 \\
c & y_2c & cx_2 & 0 & 0 & 0 & 0 & 0 & 0 \\
0 & 0 & 0 & 1 & x_2 & y_2 & c & y_2c & cx_2 \\
0 & 0 & 0 & 0 & 0 & 0 & 0 & 0 & 0 \\
c & y_3c & 0 & 0 & 0 & 0 & 0 & 0 & 0 \\
0 & 0 & 0 & 1 & 0 & y_3 & c & y_3c & 0
\end{bmatrix}
\begin{bmatrix}
c_1 \\ c_2 \\ c_3 \\ c_4 \\ c_5 \\ c_6 \\ c_7 \\ c_8 \\ c_9 \\ c_{10} \\ c_{11} \\ c_{12} \\ c_{13} \\ c_{14} \\ c_{15} \\ c_{16} \\ c_{17} \\ c_{18}
\end{bmatrix}$$

$$(9.9)$$

Symbolically the preceding equation can be expressed as

$$\mathbf{u} = \mathbf{C}_9 \mathbf{c} \qquad (9.10)$$

where $\mathbf{C}_9$ is the 18×18 submatrix in Eq. (9.9) and

$$\mathbf{u} = \{u_1 \; u_2 \; \cdots \; u_{18}\} \qquad (9.11)$$

$$\mathbf{c} = \{c_1 \; c_2 \; \cdots \; u_{18}\} \qquad (9.12)$$

Hence

$$\mathbf{c} = \mathbf{C}_9^{-1} \mathbf{u} \qquad (9.13)$$

where it can be shown that

$$\mathbf{C}_9^{-1} = \begin{bmatrix}
1 & 0 & 0 & 0 & 0 & 0 & 0 & 0 & 0 \\
0 & 0 & 0 & 0 & 0 & 0 & 1 & 0 & 0 \\
0 & 0 & 0 & 0 & 0 & 0 & 0 & 0 & 0 \\
1 & x_2 & y_2 & 0 & 0 & 0 & 0 & 0 & 0 \\
0 & 0 & 0 & 0 & 0 & 0 & 1 & x_2 & y_2 \\
0 & 0 & 0 & 0 & 0 & 0 & 0 & 0 & 0 \\
1 & x_3 & y_3 & 0 & 0 & 0 & 0 & 0 & 0 \\
0 & 0 & 0 & 0 & 0 & 0 & 1 & x_3 & y_3 \\
0 & 0 & 0 & 0 & 0 & 0 & 0 & 0 & 0 \\
1 & 0 & 0 & c & 0 & 0 & 0 & 0 & 0 \\
0 & 0 & 0 & 0 & 0 & 0 & 1 & 0 & 0 \\
0 & 0 & 0 & 0 & 0 & 0 & 0 & 0 & 0 \\
1 & x_2 & y_2 & c & y_3c & cx_2 & 0 & 0 & 0 \\
0 & 0 & 0 & 0 & 0 & 0 & 1 & x_2 & y_2 \\
0 & 0 & 0 & 0 & 0 & 0 & 0 & 0 & 0 \\
1 & x_3 & y_3 & c & y_3c & cx_3 & 0 & 0 & 0 \\
0 & 0 & 0 & 0 & 0 & 0 & 1 & x_3 & y_3 \\
0 & 0 & 0 & 0 & 0 & 0 & 0 & 0 & 0
\end{bmatrix}$$

$$
\begin{bmatrix}
0 & 0 & 0 & 0 & 0 & 0 & 0 & 0 & 0 \\
0 & 0 & 0 & 0 & 0 & 0 & 0 & 0 & 0 \\
0 & 0 & 0 & 1 & 0 & 0 & 0 & 0 & 0 \\
0 & 0 & 0 & 0 & 0 & 0 & 0 & 0 & 0 \\
0 & 0 & 0 & 0 & 0 & 0 & 0 & 0 & 0 \\
0 & 0 & 0 & 1 & x_2 & y_2 & 0 & 0 & 0 \\
0 & 0 & 0 & 0 & 0 & 0 & 0 & 0 & 0 \\
0 & 0 & 0 & 0 & 0 & 0 & 0 & 0 & 0 \\
0 & 0 & 0 & 1 & x_3 & y_3 & 0 & 0 & 0 \\
0 & 0 & 0 & 0 & 0 & 0 & 0 & 0 & 0 \\
c & 0 & 0 & 0 & 0 & 0 & 0 & 0 & 0 \\
0 & 0 & 0 & 1 & 0 & 0 & c & 0 & 0 \\
0 & 0 & 0 & 0 & 0 & 0 & 0 & 0 & 0 \\
c & y_2c & cx_2 & 0 & 0 & 0 & 0 & 0 & 0 \\
0 & 0 & 0 & 1 & x_2 & y_2 & c & y_2c & cx_2 \\
0 & 0 & 0 & 0 & 0 & 0 & 0 & 0 & 0 \\
c & y_3c & cx_3 & 0 & 0 & 0 & 0 & 0 & 0 \\
0 & 0 & 0 & 1 & x_3 & y_3 & c & y_3c & cx_3
\end{bmatrix}
\tag{9.14}
$$

The strains in the element are computed from

$$
\mathbf{e} =
\begin{bmatrix}
e_{xx} \\ e_{yy} \\ e_{zz} \\ e_{xy} \\ e_{yz} \\ e_{zx}
\end{bmatrix}
=
\begin{bmatrix}
\partial u/\partial x \\
\partial v/\partial y \\
\partial w/\partial z \\
\partial v/\partial x + \partial u/\partial y \\
\partial w/\partial y + \partial v/\partial z \\
\partial u/\partial z + \partial w/\partial x
\end{bmatrix}
$$

$$
=
\begin{bmatrix}
0 & 1 & 0 & 0 & 0 & z & 0 & 0 & 0 & 0 & 0 & 0 & 0 & 0 & 0 & 0 & 0 & 0 \\
0 & 0 & 0 & 0 & 0 & 0 & 0 & 0 & 1 & 0 & z & 0 & 0 & 0 & 0 & 0 & 0 & 0 \\
0 & 0 & 0 & 0 & 0 & 0 & 0 & 0 & 0 & 0 & 0 & 0 & 0 & 0 & 1 & y & x & 0 \\
0 & 0 & 1 & 0 & z & 0 & 0 & 1 & 0 & 0 & 0 & z & 0 & 0 & 0 & 0 & 0 & 0 \\
0 & 0 & 0 & 0 & 0 & 0 & 0 & 0 & 1 & y & x & 0 & 0 & 1 & 0 & z & 0 \\
0 & 0 & 0 & 1 & y & x & 0 & 0 & 0 & 0 & 0 & 1 & 0 & 0 & 0 & z
\end{bmatrix}
\mathbf{C}_9^{-1}\mathbf{u}
\tag{9.15}
$$

The preceding equation can be rewritten as

$$
\mathbf{e} = \mathbf{b}_9\mathbf{u} = \mathbf{D}_9\mathbf{C}_9^{-1}\mathbf{u}
\tag{9.16}
$$

where

$$
\mathbf{b}_9 = \mathbf{D}_9\mathbf{C}_9^{-1}
\tag{9.17}
$$

and $\mathbf{D}_9$ is the 6×18 matrix in Eq. (9.15). Subsequently, the stiffness matrix can be computed from

$$
\mathbf{k}_9 = \iiint \mathbf{b}_9^T \mathbf{E}_3 \mathbf{b}_9 \, dx\, dy\, dz
$$

$$
= (\mathbf{C}_9^{-1})^T \iiint \mathbf{D}_9^T \mathbf{E}_3 \mathbf{D}_9 \, dx\, dy\, dz \, \mathbf{C}_9^{-1}
\tag{9.18}
$$

The integration with respect to z presents no difficulty because the range of integration is from 0 to c; however, further integration with respect to x and y can be best accomplished by changing the rectangular coordinates x and y into triangular coordinates ξ and η as shown in Fig. 9.2 for which the range of integration is from 0 to 1.

The rectangular coordinates x and y are expressed in terms of the triangular coordinates ξ and η by

$$x = \xi x_2 (1 - \eta) \tag{9.19}$$

$$y = \xi (y_2 - \eta y_{23}) \tag{9.20}$$

where

$$y_{23} = y_2 - y_3 \tag{9.21}$$

Also the coordinate z can be expressed in terms of the nondimensional coordinate ζ as

$$z = c\zeta \tag{9.22}$$

Using Eqs. (9.19), (9.20), and (9.22), it can be shown that the determinant of the Jacobian $J(x, y, z)$ is given by

$$|J(x, y, z)| = \begin{vmatrix} \partial x/\partial \xi & \partial x/\partial \eta & \partial x/\partial \zeta \\ \partial y/\partial \xi & \partial y/\partial \eta & \partial y/\partial \zeta \\ \partial z/\partial \xi & \partial z/\partial \eta & \partial z/\partial \zeta \end{vmatrix} = -2c\xi A_{123} \tag{9.23}$$

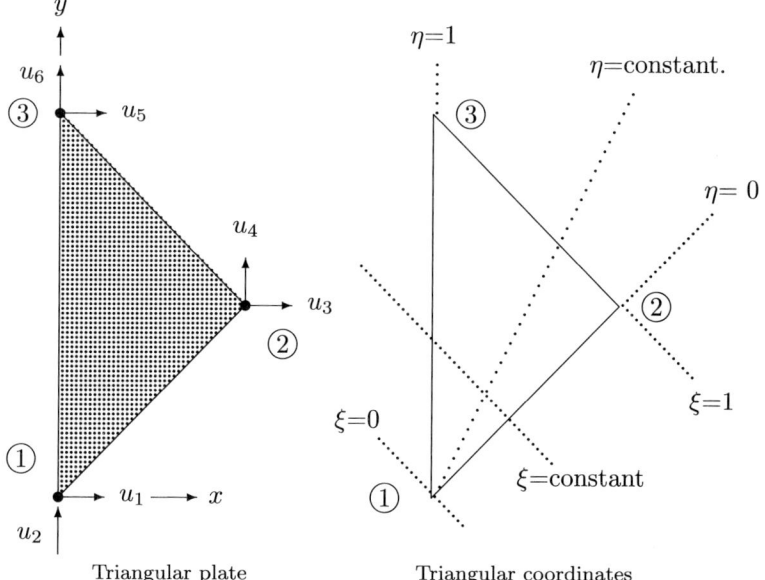

Triangular plate   Triangular coordinates

Fig. 9.2   Node displacements and triangular coordinates.

where A_{123} is the area of the triangle with vertices 1, 2, 3. The element stiffness matrix $\mathbf{k}_9$ can therefore be calculated from the modified Eq. (9.18) as

$$
\begin{aligned}
\mathbf{k}_9 &= (\mathbf{C}_9^{-1})^T \int_0^1 \int_0^1 \int_0^1 \mathbf{D}_9^T(\xi, \eta, \eta) \mathbf{E}_3 \mathbf{D}_9 |J(x, y, z)| \, d\xi \, d\eta \, d\zeta \, \mathbf{C}_9^{-1} \\
&= 2cA_{123} \mathbf{C}_9^{-1} \int_0^1 \int_0^1 \int_0^1 \mathbf{D}_9^T(\xi, \eta, \zeta) \mathbf{E}_3 \mathbf{D}_9(\xi, \eta, \zeta) \xi \, d\xi \, d\eta \, d\zeta \, \mathbf{C}_9^{-1}
\end{aligned} \qquad (9.24)
$$

$\mathbf{C}_9$ must be expressed in terms of the ξ, η, and ζ coordinates.

Thermal Stiffness

The thermal stiffness $\mathbf{h}_9$ is determined from

$$
\begin{aligned}
\mathbf{h}_9 &= \int_v \frac{E\mathbf{b}_9^T}{(1 - 2v)} \begin{bmatrix} -1 \\ -1 \\ -1 \\ 0 \\ 0 \\ 0 \end{bmatrix} dV = \iiint \mathbf{H}_9 \, dx \, dy \, dz \\
&= \iiint \mathbf{H}_9(\xi, \eta, \zeta) |J(x, y, z)| \, d\xi \, d\eta \, d\zeta \\
&= 2cA_{123} \int_0^1 \int_0^1 \int_0^1 \mathbf{H}_9(\xi, \eta, \zeta) \xi \, d\xi \, d\eta \, d\zeta
\end{aligned} \qquad (9.25)
$$

where

$$
\mathbf{H}_9 = \frac{E\mathbf{b}_9^T}{(1 - 2v)} \begin{bmatrix} -1 \\ -1 \\ -1 \\ 0 \\ 0 \\ 0 \end{bmatrix} \qquad (9.26)
$$

Thermal Loads

The thermal loads $\mathbf{q}_9$ are determined from

$$
\mathbf{q}_9 = \alpha T \mathbf{h}_9 \qquad (9.27)
$$

Mass Matrix

The displacements u, v, and w within the pentahedron are obtained from Eq. (9.1) as

$$\begin{bmatrix} u \\ v \\ w \end{bmatrix} = \begin{bmatrix} 1 & x & y & z & yz & zx & 0 & 0 & 0 & 0 & 0 & 0 & 0 & 0 & 0 & 0 & 0 & 0 \\ 0 & 0 & 0 & 0 & 0 & 0 & 1 & x & y & z & yz & zx & 0 & 0 & 0 & 0 & 0 & 0 \\ 0 & 0 & 0 & 0 & 0 & 0 & 0 & 0 & 0 & 0 & 0 & 0 & 1 & x & y & z & yz & zx \end{bmatrix} \begin{bmatrix} c_1 \\ \vdots \\ c_{18} \end{bmatrix}$$

$$= \mathbf{A}_9 \mathbf{c} = \mathbf{A}_9 \mathbf{C}_9^{-1} \mathbf{u} \tag{9.28}$$

where $\mathbf{A}_9$ is the 3×18 matrix in Eq. (9.28) and

$$\mathbf{a}_9 = \mathbf{A}_9 \mathbf{C}_9^{-1} \tag{9.29}$$

The mass matrix $\mathbf{m}_{13}$ is then determined from

$$\mathbf{m}_9 = \rho \int_v \mathbf{a}_9^T \mathbf{a}_9 \, dV = \rho \iiint \mathbf{a}_9^T \mathbf{a}_9 \, dx \, dy \, dz$$

$$= \rho \int_0^1 \int_0^1 \int_0^1 \mathbf{a}_9^T \mathbf{a}_9 |J(x,y,z)| \, d\xi \, d\eta \, d\zeta$$

$$= 2cA_{123}\rho \int_0^1 \int_0^1 \int_0^1 \mathbf{a}_9^T \mathbf{a}_9 \xi \, d\xi \, d\eta \, d\zeta \tag{9.30}$$

where $\mathbf{a}_9$ is expressed in terms of the nondimensional cordinates ξ, η, and ζ.

12
Prismatic Pentahedron T10: Assumed Displacement Distribution plus Corrective Distribution Inside the Element Boundaries

The element displacements for a pentahedron element are shown in Fig. 10.1. The element forces (not shown in this figure) are in the same directions as the corresponding element displacements. The origin of the rectangular coordinate system x, y, z is assumed to be at the node (1) with *the Oy axis coinciding with the element edge 1–3*. The nondimensional coordinate system (triangular non-dimensional coordinate system in the x–y plane) and the rectangular coordinate system are shown in Fig. 10.2. The triangular nondimensional system is needed here to perform integration within the element boundaries.

The x and y coordinates are functions of the ξ and η triangular coordinates as shown here:

$$x = \xi(1 - \eta)x_2 \tag{10.1}$$

$$y = \xi(y_2 - \eta y_{23}) \tag{10.2}$$

where

$$y_{23} = y_2 - y_3 \tag{10.3}$$

and

$$z = \zeta c \tag{10.4}$$

where c is the length of the prismatic pentahedron. This leads then to the formulas for ξ, η, and ζ as

$$\xi = \frac{x}{x_2(1 - \eta)} \tag{10.5}$$

$$\eta = \frac{1}{\xi y_{23}}(\xi y_2 - y) \tag{10.6}$$

$$\zeta = \frac{z}{c} \tag{10.7}$$

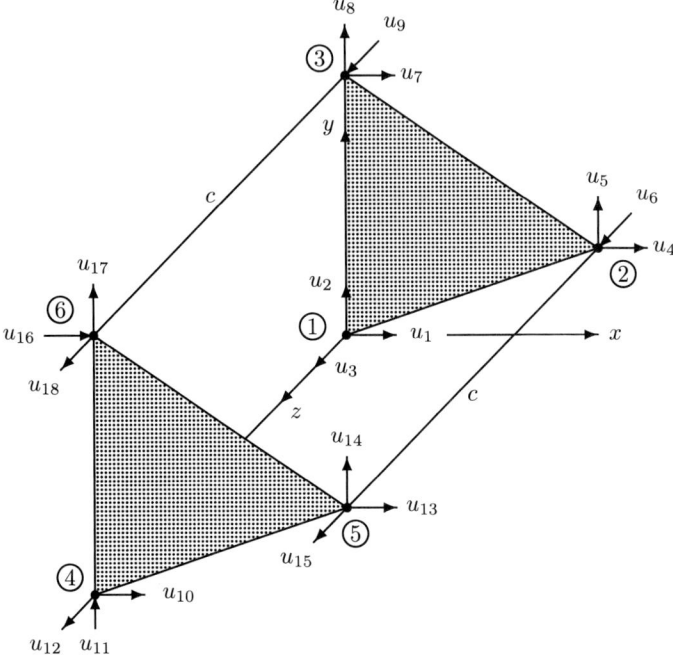

Fig. 10.1   Node displacements for prismatic pentahedron. (The shaded sides are in the *x–y* planes.)

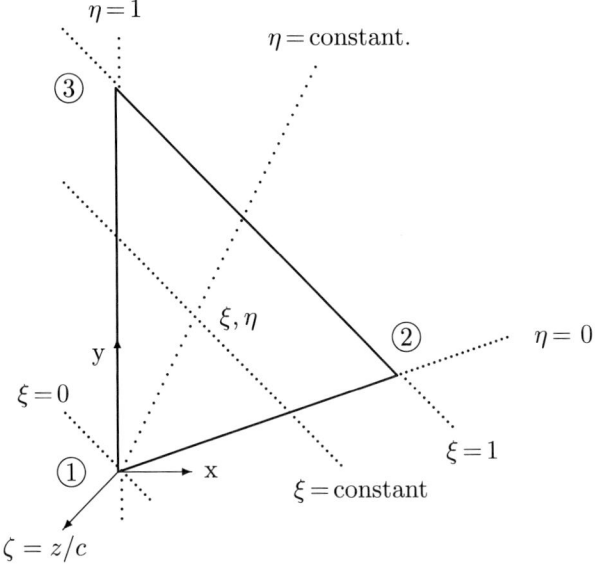

Fig. 10.2   Triangular coordinates.

The displacements within the element u, v, and w in the x, y, and z directions are assumed to be given by

$$u = c_1 + c_2 x + c_3 y + c_4 z + c_5 xy + c_6 zx$$
$$+ c_{19} f(\xi, \eta, \zeta) \tag{10.8}$$
$$v = c_7 + c_8 x + c_9 y + c_{10} z + c_{11} yz + c_{12} xy$$
$$+ c_{20} f(\xi, \eta, \zeta) \tag{10.9}$$
$$w = c_{13} + c_{14} x + c_{15} y + c_{16} z + c_{17} zx + c_{18} yz$$
$$+ c_{21} f(\xi, \eta, \zeta) \tag{10.10}$$

These assumed displacements ensure that all three direct strains e_{xx}, e_{yy}, and e_{zz} vary linearly in cross-wise directions. The selected function $f(\xi, \eta, \zeta)$ vanishes on the element boundaries and can be assumed as

$$f(\xi, \eta, \zeta) = \xi(1 - \xi)\eta(1 - \eta)\zeta(1 - \zeta) \tag{10.11}$$

For the selected coordinate system it follows that

$$x_1 = x_3 = x_4 = x_6 = 0 \tag{10.12}$$
$$y_1 = y_4 = 0 \tag{10.13}$$
$$z_1 = z_2 = z_3 = 0 \tag{10.14}$$

Because the shaded triangular planes shown in Fig. 10.1 are parallel to each other and perpendicular to the z axis,

$$x_5 = x_2 \tag{10.15}$$
$$y_5 = y_2 \tag{10.16}$$
$$y_6 = y_3 \tag{10.17}$$
$$z_4 = z_5 = z_6 = c \tag{10.18}$$

where c is the length of the prismatic pentahedron. Equations (10.8–10.10) can be represented as

$$\begin{bmatrix} u \\ v \\ w \end{bmatrix} = [\mathbf{G}_a \ \mathbf{G}_b]\mathbf{c} = \mathbf{Gc} \tag{10.19}$$

where $\mathbf{G}$ is a rectangular matrix whose elements are the assumed displacement functions. The matrix $\mathbf{G}$ consists of two submatrices $\mathbf{G}_a$ and $\mathbf{G}_b$ given by

$$\mathbf{G}_a = \begin{bmatrix} 1 & x & y & z & xy & zx & 0 & 0 & 0 & 0 & 0 & 0 & 0 & 0 & 0 & 0 & 0 & 0 \\ 0 & 0 & 0 & 0 & 0 & 0 & 1 & x & y & z & yz & xy & 0 & 0 & 0 & 0 & 0 & 0 \\ 0 & 0 & 0 & 0 & 0 & 0 & 0 & 0 & 0 & 0 & 0 & 0 & 1 & x & y & z & zx & yz \end{bmatrix} \tag{10.20}$$

and

$$\mathbf{G}_b = \begin{bmatrix} f(\xi,\eta,\zeta) \\ f(\xi,\eta,\zeta) \\ f(\xi,\beta,\zeta) \end{bmatrix} \tag{10.21}$$

while the column matrix $\mathbf{c}$ is given by

$$\mathbf{c} = \{c_1 c_2 \cdots c_{21}\} = \{\mathbf{c}_a\ \mathbf{c}_b\} \tag{10.22}$$

where

$$\mathbf{c}_a = \{c_1 \cdots c_{18}\} \tag{10.23}$$

$$\mathbf{c}_b = \{c_{19}\ c_{20}\ c_{21}\} \tag{10.24}$$

Thus using Eqs. (10.7–10.9) together with the coordinate relations [Eqs. (10.12–10.18)], the element node displacements can be expressed as

$$\mathbf{u} = \begin{bmatrix} u_1 \\ u_2 \\ u_3 \\ u_4 \\ u_5 \\ u_6 \\ u_7 \\ u_8 \\ u_9 \\ u_{10} \\ u_{11} \\ u_{12} \\ u_{13} \\ u_{14} \\ u_{15} \\ u_{16} \\ u_{17} \\ u_{18} \end{bmatrix} = \begin{bmatrix} 1 & 0 & 0 & 0 & 0 & 0 & 0 & 0 & 0 & 0 & 0 & 0 \\ 0 & 0 & 0 & 0 & 0 & 0 & 1 & 0 & 0 & 0 & 0 & 0 \\ 0 & 0 & 0 & 0 & 0 & 0 & 0 & 0 & 0 & 0 & 0 & 0 \\ 1 & x_2 & 0 & 0 & 0 & 0 & 0 & 0 & 0 & 0 & 0 & 0 \\ 0 & 0 & 0 & 0 & 0 & 0 & 1 & x_2 & y_2 & 0 & 0 & 0 \\ 0 & 0 & 0 & 0 & 0 & 0 & 0 & 0 & 0 & 0 & 0 & 0 \\ 1 & 0 & y_3 & 0 & 0 & 0 & 0 & 0 & 0 & 0 & 0 & 0 \\ 0 & 0 & 0 & 0 & 0 & 0 & 1 & 0 & y_3 & 0 & 0 & 0 \\ 0 & 0 & 0 & 0 & 0 & 0 & 0 & 0 & 0 & 0 & 0 & 0 \\ 1 & 0 & 0 & c & 0 & 0 & 0 & 0 & 0 & 0 & 0 & 0 \\ 0 & 0 & 0 & 0 & 0 & 0 & 1 & 0 & 0 & c & 0 & 0 \\ 0 & 0 & 0 & 0 & 0 & 0 & 0 & 0 & 0 & 0 & 0 & 0 \\ 1 & x_2 & y_2 & c & x_2y_2 & cx_2 & 0 & 0 & 0 & 0 & 0 & 0 \\ 0 & 0 & 0 & 0 & 0 & 0 & 1 & x_2 & y_2 & c & y_2c & x_2y_2 \\ 0 & 0 & 0 & 0 & 0 & 0 & 0 & 0 & 0 & 0 & 0 & 0 \\ 1 & 0 & y_3 & c & y_3c & 0 & 0 & 0 & 0 & 0 & 0 & 0 \\ 0 & 0 & 0 & 0 & 0 & 0 & 1 & 0 & y_3 & c & y_3c & 0 \\ 0 & 0 & 0 & 0 & 0 & 0 & 0 & 0 & 0 & 0 & 0 & 0 \end{bmatrix}$$

$$\begin{bmatrix} 0 & 0 & 0 & 0 & 0 & 0 \\ 0 & 0 & 0 & 0 & 0 & 0 \\ 1 & 0 & 0 & 0 & 0 & 0 \\ 0 & 0 & 0 & 0 & 0 & 0 \\ 0 & 0 & 0 & 0 & 0 & 0 \\ 1 & x_2 & y_2 & 0 & 0 & 0 \\ 0 & 0 & 0 & 0 & 0 & 0 \\ 0 & 0 & 0 & 0 & 0 & 0 \\ 1 & 0 & y_3 & 0 & 0 & 0 \\ 0 & 0 & 0 & 0 & 0 & 0 \\ 0 & 0 & 0 & 0 & 0 & 0 \\ 1 & 0 & 0 & c & 0 & 0 \\ 0 & 0 & 0 & 0 & 0 & 0 \\ 0 & 0 & 0 & 0 & 0 & 0 \\ 1 & x_2 & y_2 & c & cx_2 & 0 \\ 0 & 0 & 0 & 0 & 0 & 0 \\ 0 & 0 & 0 & 0 & 0 & 0 \\ 1 & 0 & y_3 & c & 0 & cy_3 \end{bmatrix} \begin{bmatrix} c_1 \\ c_2 \\ c_3 \\ c_4 \\ c_5 \\ c_6 \\ c_7 \\ c_8 \\ c_9 \\ c_{10} \\ c_{11} \\ c_{12} \\ c_{13} \\ c_{14} \\ c_{15} \\ c_{16} \\ c_{17} \\ c_{18} \end{bmatrix}$$

(10.25)

The preceding equation can be expressed symbolically as

$$\mathbf{u} = \mathbf{C}_{10}\mathbf{c}_a \tag{10.26}$$

where

$$\mathbf{u} = \{u_1 \ u_2 \cdots u_{18}\} \tag{10.27}$$

and $\mathbf{C}_{13a}$ is the 18×18 submatrix in Eq. (10.25). Hence,

$$\mathbf{c} = \mathbf{C}_{10}^{-1}\mathbf{u} \tag{10.28}$$

To obtain the element strains, the partial derivatives of the displacements must be obtained first. To find the partial derivatives of the function $f(\xi, \eta, \zeta)$, the chain rule of differentiation shown next is used:

$$\frac{\partial}{\partial x} = \frac{\partial}{\partial \xi}\frac{\partial \xi}{\partial x} + \frac{\partial}{\partial \eta}\frac{\partial \eta}{\partial x} + \frac{\partial}{\partial \zeta}\frac{\partial \zeta}{\partial x} \tag{10.29}$$

$$\frac{\partial}{\partial y} = \frac{\partial}{\partial \xi}\frac{\partial \xi}{\partial y} + \frac{\partial}{\partial \eta}\frac{\partial \eta}{\partial y} + \frac{\partial}{\partial \zeta}\frac{\partial \zeta}{\partial y} \tag{10.30}$$

$$\frac{\partial}{\partial z} = \frac{\partial}{\partial \xi}\frac{\partial \xi}{\partial z} + \frac{\partial}{\partial \eta}\frac{\partial \eta}{\partial z} + \frac{\partial}{\partial \zeta}\frac{\partial \zeta}{\partial z} \tag{10.31}$$

From the relationship between the rectangular and triangular coordinates in Eqs. (10.5–10.7), it follows that

$$\frac{\partial \xi}{\partial x} = \frac{1}{x_2(1-\eta)} \tag{10.32}$$

$$\frac{\partial \eta}{\partial x} = 0 \tag{10.33}$$

$$\frac{\partial \zeta}{\partial x} = 0 \tag{10.34}$$

$$\frac{\partial \xi}{\partial y} = 0 \tag{10.35}$$

$$\frac{\partial \eta}{\partial y} = -\frac{1}{\xi y_{23}} \tag{10.36}$$

$$\frac{\partial \zeta}{\partial y} = 0 \tag{10.37}$$

$$\frac{\partial \xi}{\partial z} = 0 \tag{10.38}$$

$$\frac{\partial \eta}{\partial z} = 0 \tag{10.39}$$

$$\frac{\partial \zeta}{\partial z} = \frac{1}{c} \tag{10.40}$$

Using now the relations in Eqs. (10.29–10.31) and Eqs. (10.32–10.40),

$$\frac{\partial f}{\partial x} = \frac{\partial f}{\partial \xi}\frac{\partial \xi}{\partial x} = (1-2\xi)\eta\zeta\frac{(1-\zeta)}{x_2} \tag{10.41}$$

$$\frac{\partial f}{\partial y} = \frac{\partial f}{\partial \eta}\frac{\partial \eta}{\partial y} = \frac{-(1-\xi)(1-2\eta)}{y_{23}} \tag{10.42}$$

$$\frac{\partial f}{\partial z} = \frac{\partial f}{\partial \zeta}\frac{\partial \zeta}{\partial z} = \frac{\xi(1-\xi)(1-\eta)(1-2\zeta)}{c} \tag{10.43}$$

The strains in the element can now be computed from

$$e_{xx} = \frac{\partial u}{\partial x} = c_2 + c_5 y + c_6 z + c_{19}(1-2\xi)\eta/\zeta(1-\zeta)x_2 \tag{10.44}$$

$$e_{yy} = \frac{\partial v}{\partial y} = c_9 + c_{11}z + c_{12}x + c_{20}(1-\xi)(1-2\eta)\zeta(1-\zeta)/y_{23} \tag{10.45}$$

$$e_{zz} = \frac{\partial w}{\partial z} = c_{16} + C_{17}X + C_{18}Y + C_{21}\xi(1-\xi)\eta(1-\eta)(1-2\zeta)/c \tag{10.46}$$

$$e_{xy} = \frac{\partial v}{\partial x} + \frac{\partial u}{\partial y} = c_3 + c_5 x + c_8 + c_{19}(1 - \xi)(1 - 2\eta)\zeta(1 - \zeta)/y_{23}$$

$$+ c_{20}(1 - 2\xi)\eta\zeta(1 - \zeta)/x_2 \qquad (10.47)$$

$$e_{yz} = \frac{\partial w}{\partial y} + \frac{\partial v}{\partial z} = c_{10} + c_{11}y + c_{18}z + c_{15} + c_{20}\xi(1 - \xi)\eta(1 - 2\zeta)/c$$

$$+ c_{21}(1 - \xi)(1 - 2\eta)\zeta(1 - \zeta)/y_{23} \qquad (10.48)$$

$$e_{zx} = \frac{\partial u}{\partial z} + \frac{\partial w}{\partial x} = c_4 + c_6 x + c_{14} + c_{17}z + c_{19}\xi(1 - \xi)\eta(1 - \eta)(1 - 2\zeta)/c$$

$$+ c_{21}(1 - 2\xi)\eta\zeta(1 - \zeta)/x_2 \qquad (10.49)$$

which leads to

$$\mathbf{e} = \begin{bmatrix} e_{xx} \\ e_{yy} \\ e_{xy} \end{bmatrix} = \mathbf{Hc} = [\mathbf{H}_a \quad \mathbf{H}_b] \begin{bmatrix} \mathbf{c}_a \\ \mathbf{c}_b \end{bmatrix} \qquad (10.50)$$

where

$$\mathbf{H} = [\mathbf{H}_a \quad \mathbf{H}_b] \qquad (10.51)$$

$$\mathbf{H}_a = \begin{bmatrix} 0 & 1 & 0 & 0 & y & z & 0 & 0 & 0 & 0 & 0 & 0 & 0 & 0 & 0 & 0 & 0 & 0 \\ 0 & 0 & 0 & 0 & 0 & 0 & 0 & 0 & 1 & 0 & z & x & 0 & 0 & 0 & 0 & 0 & 0 \\ 0 & 0 & 0 & 0 & 0 & 0 & 0 & 0 & 0 & 0 & 0 & 0 & 0 & 0 & 0 & 1 & y & 0 \\ 0 & 0 & 1 & 0 & x & 0 & 0 & 1 & 0 & 0 & 0 & 0 & 0 & 0 & 0 & 0 & 0 & 0 \\ 0 & 0 & 0 & 0 & 0 & 0 & 0 & 0 & 0 & 1 & y & 0 & 0 & 0 & 1 & 0 & 0 & z \\ 0 & 0 & 0 & 1 & 0 & x & 0 & 0 & 0 & 0 & 0 & 0 & 0 & 1 & 0 & 0 & z & 0 \end{bmatrix}$$

$$(10.52)$$

$$\mathbf{H}_b = \begin{bmatrix} (1 - \xi)\eta\zeta(1 - \zeta)/x_2 & 0 \\ 0 & -(1 - \xi)(1 - 2\eta)\zeta(1 - \zeta)/y_{23} \\ 0 & 0 \\ (1 - \xi)(1 - 2\eta)\zeta(1 - \zeta)/y_{23} & (1 - 2\xi)\eta\zeta(1 - \zeta)/x_2 \\ 0 & (1 - \xi)\eta(1 - \eta)(1 - 2\zeta)/c \\ \xi\eta(1 - \eta)(1 - 2\zeta)/c & 0 \end{bmatrix}$$

$$\begin{bmatrix} 0 \\ 0 \\ \xi(1 - \xi)\eta(1 - \eta)(1 - 2\zeta)/c \\ 0 \\ -(1 - \xi)(1 - 2\eta)\zeta(1 - \zeta)/y_{23} \\ (1 - 2\xi)\eta\zeta(1 - \zeta)/x_2 \end{bmatrix} \qquad (10.53)$$

Combining now Eq. (10.28) with the identity $\mathbf{c}_b = \mathbf{c}_b$ yields

$$
\mathbf{c} = \begin{bmatrix} \mathbf{c}_a \\ \mathbf{c}_b \end{bmatrix} = \begin{bmatrix} \mathbf{C}_a^{-1} & \mathbf{0} \\ \mathbf{0} & \mathbf{I} \end{bmatrix} \begin{bmatrix} \mathbf{u} \\ \mathbf{c}_b \end{bmatrix} = \mathbf{W}\hat{\mathbf{u}}
\tag{10.54}
$$

where

$$
\mathbf{W} = \begin{bmatrix} \mathbf{C}_a^{-1} & \mathbf{0} \\ \mathbf{0} & \mathbf{I} \end{bmatrix}
\tag{10.55}
$$

and

$$
\hat{\mathbf{u}} = \begin{bmatrix} \mathbf{u} \\ \mathbf{c}_b \end{bmatrix}
\tag{10.56}
$$

The strain energy U_i in the element is given by

$$
U_i = \frac{1}{2} \int_v \mathbf{e}^T \mathbf{E}_3 \mathbf{e}^T \, dV = \frac{1}{2} \mathbf{c}^T \int_v \mathbf{H}^T \mathbf{E}_3 \mathbf{H} \, dV \mathbf{c} = \frac{1}{2} \hat{\mathbf{u}}^T \mathbf{W}^T \int_v \mathbf{H}^T \mathbf{E}_3 \mathbf{H} \, dV \, \mathbf{W}\hat{\mathbf{u}}
$$

$$
= \frac{1}{2} \hat{\mathbf{u}}^T \hat{\mathbf{k}} \hat{\mathbf{u}}
\tag{10.57}
$$

where

$$
\hat{\mathbf{k}} = \mathbf{W}^T \int_v \mathbf{H}^T \mathbf{E}_3 \mathbf{H} \, dV \, \mathbf{W} = \begin{bmatrix} \mathbf{k}_{aa} & \mathbf{k}_{ab} \\ \mathbf{k}_{ba} & \mathbf{k}_{bb} \end{bmatrix}
\tag{10.58}
$$

and the expressions for the three-dimensional Young's modulus is given by

$$
\mathbf{E}_3 = \frac{E}{(1+v)(1-2v)}
$$

$$
\times \begin{bmatrix}
(1-v) & v & v & 0 & 0 & 0 \\
v & (1-v) & v & 0 & 0 & 0 \\
v & v & (1-v) & 0 & 0 & 0 \\
0 & 0 & 0 & (1-2v) & 0 & 0 \\
0 & 0 & 0 & 0 & (1-2v) & 0 \\
0 & 0 & 0 & 0 & 0 & (1-2v)
\end{bmatrix}
\tag{10.59}
$$

The volume integral in Eq. (10.58) is calculated from

$$
\int_v \mathbf{H}^T \mathbf{E}_3 \mathbf{H} \, dV = \iiint \mathbf{H}^T \mathbf{E}_3 \mathbf{H} \, dx \, dy \, dz
$$

$$
= \int_{\zeta=0}^{\zeta=1} \int_{\eta=0}^{\eta=1} \int_{\xi=0}^{\xi=1} \mathbf{H}^T \mathbf{E}_3 \mathbf{H}^T |J(x,y,z)| \, d\xi \, d\eta \, d\zeta
\tag{10.60}
$$

where $|J(x, y, z)|$ is the determinant of the Jacobian J given by

$$J = |J(x, y, z)| = \begin{vmatrix} \dfrac{\partial x}{\partial \xi} & \dfrac{\partial x}{\partial \eta} & \dfrac{\partial x}{\partial \zeta} \\[2mm] \dfrac{\partial y}{\partial \xi} & \dfrac{\partial y}{\partial \eta} & \dfrac{\partial y}{\partial \zeta} \\[2mm] \dfrac{\partial z}{\partial \xi} & \dfrac{\partial z}{\partial \eta} & \dfrac{\partial z}{\partial \zeta} \end{vmatrix} = \begin{vmatrix} (1 - \eta)x_2 & -\xi_2 & 0 \\ (y_2 - \eta)y_{23} & -\xi y_{23} & 0 \\ 0 & 0 & c \end{vmatrix} = -\xi x_2 y_3 c$$

(10.61)

For a volume element $dx\,dy\,dz = |J|\,d\xi\,d\eta\,d\zeta$, and also in this case $x_2 y_3 c = 2\times$ element volume.

The total potential energy U in the element can be written as

$$U = U_i - \mathbf{u}^T \mathbf{S} = \frac{1}{2}\hat{\mathbf{u}}^T \hat{\mathbf{k}}\hat{\mathbf{u}} - \mathbf{u}^T \mathbf{S} = \frac{1}{2}[\mathbf{u}^T \ \mathbf{c}_b^T]\begin{bmatrix} \mathbf{k}_{aa} & \mathbf{k}_{ab} \\ \mathbf{k}_{ba} & \mathbf{k}_{bb} \end{bmatrix}\begin{bmatrix} \mathbf{u} \\ \mathbf{c}_b \end{bmatrix} - [\mathbf{u}^T \ \mathbf{c}_b^T]\begin{bmatrix} \mathbf{S} \\ \mathbf{0} \end{bmatrix}$$

(10.62)

where $\mathbf{S}$ is a column matrix of the element forces corresponding with the displacements $\mathbf{u}$ and the term $\mathbf{u}^T \mathbf{S}$ is the potential of external forces. Now the condition of minimum potential energy requires that

$$\frac{\partial U}{\partial \hat{\mathbf{u}}} = \mathbf{0}$$

(10.63)

leading to

$$\begin{bmatrix} \mathbf{k}_{aa} & \mathbf{k}_{ab} \\ \mathbf{k}_{ba} & \mathbf{k}_{bb} \end{bmatrix}\begin{bmatrix} \mathbf{u} \\ \mathbf{c}_b \end{bmatrix} - \begin{bmatrix} \mathbf{S} \\ \mathbf{0} \end{bmatrix} = \begin{bmatrix} \mathbf{0} \\ \mathbf{0} \end{bmatrix}$$

(10.64)

The preceding expression can be derived by multiplying first the individual matrices in U and then performing partial differentiation with respect to $u_1, u_2, \ldots$ and then condensing the resulting equations into the matrix equation Eq. (10.64).

The matrix $\mathbf{c}_b$ can then be calculated from the second row in the preceding equation as

$$\mathbf{c}_b = -\mathbf{k}_{bb}^{-1}\mathbf{k}_{ba}\mathbf{u}$$

(10.65)

which when substituted into the first row in Eq. (10.64) results in

$$\mathbf{k}_{10} - \mathbf{k}_{ab}\mathbf{k}_{bb}^{-1}\mathbf{k}_{ba}\mathbf{u} = \mathbf{S}$$

(10.66)

Hence, by definition, the element stiffness matrix $\mathbf{k}_{10}$ is given by

$$\mathbf{k}_{10} = \mathbf{k}_{aa} - \mathbf{k}_{ab}\mathbf{k}_{bb}^{-1}\mathbf{k}_{ba}$$

(10.67)

The strains $\mathbf{e}$ are now determined from Eq. (10.32), which can be rewritten as

$$\mathbf{e} = \mathbf{H}_a \mathbf{c}_a + \mathbf{H}_b \mathbf{c}_b$$

(10.68)

Substituting Eq. (10.28) and Eq. (10.65) into Eq. (10.68), the strains $\mathbf{e}$ are given by

$$\mathbf{e} = (\mathbf{H}_a \mathbf{C}_a^{-1} - \mathbf{H}_b \mathbf{k}_{bb}^{-1})\mathbf{u} \tag{10.69}$$

The corresponding stresses are then obtained from

$$\sigma = \mathbf{E}_3 \mathbf{e} \tag{10.70}$$

$$= (\mathbf{C}_{13a}^{-1})^T \iiint \mathbf{D}_{14}^T \mathbf{E}_3 \mathbf{D}_{10} \, dx \, dy \, dz \, \mathbf{C}_{13}^{-1} \tag{10.71}$$

Thermal Stiffness

The thermal stiffness $\mathbf{h}_{10}$ is determined from

$$\mathbf{h}_{10} = \int_v \frac{E\mathbf{b}_{10}^T}{(1-2v)} \begin{bmatrix} -1 \\ -1 \\ -1 \\ 0 \\ 0 \\ 0 \end{bmatrix} dV = \iiint \mathbf{H}_{10} \, dx \, dy \, dz$$

$$= \iiint \mathbf{H}_{10}(\xi, \eta, \zeta)|J(x, y, z)| \, d\xi \, d\eta \, d\zeta$$

$$= 2cA_{123} \int_0^1 \int_0^1 \int_0^1 \mathbf{H}_{10}(\xi, \eta, \zeta)\xi \, d\xi \, d\eta \, d\zeta \tag{10.72}$$

where

$$\mathbf{H}_{10} = \frac{E\mathbf{b}_{10}^T}{(1-2v)} \begin{bmatrix} -1 \\ -1 \\ -1 \\ 0 \\ 0 \\ 0 \end{bmatrix} \tag{10.73}$$

Thermal Loads

The thermal loads $\mathbf{q}_{14}$ are determined from

$$\mathbf{q}_{10} = \alpha T \mathbf{h}_{10} \tag{10.74}$$

Mass Matrix

Neglecting the corrective displacement field within the element boundaries, the displacements u, v, and w within the pentahedron are obtained from

Eq. (10.20) as

$$
\begin{bmatrix} u \\ v \\ w \end{bmatrix} = \begin{bmatrix} 1 & x & y & z & yz & zx & 0 & 0 & 0 & 0 & 0 & 0 & 0 & 0 & 0 & 0 & 0 & 0 \\ 0 & 0 & 0 & 0 & 0 & 0 & 1 & x & y & z & yz & zx & 0 & 0 & 0 & 0 & 0 & 0 \\ 0 & 0 & 0 & 0 & 0 & 0 & 0 & 0 & 0 & 0 & 0 & 0 & 1 & x & y & z & yz & zx \end{bmatrix}
$$

$$
\times \begin{bmatrix} c_1 \\ \vdots \\ c_{18} \end{bmatrix} = \mathbf{G}_a \mathbf{c}_a = \mathbf{a}_{10} \mathbf{u} \tag{10.75}
$$

where

$$
\mathbf{a}_{10} = \mathbf{G}_a \mathbf{C}_{10}^{-1} \tag{10.76}
$$

The mass matrix $\mathbf{m}_{13a}$ is then determined from

$$
\mathbf{m}_{10} = \rho \int_v \mathbf{a}_{10}^T \mathbf{a}_{10} \, \mathrm{d}V = \rho \iiint \mathbf{a}_{10}^T \mathbf{a}_{10} \, \mathrm{d}x \, \mathrm{d}y \, \mathrm{d}z
$$

$$
= \rho \int_0^1 \int_0^1 \int_0^1 \mathbf{a}_{10}^T \mathbf{a}_{10} |J(x, y, z)| \, \mathrm{d}\xi \, \mathrm{d}\eta \, \mathrm{d}\zeta
$$

$$
= 2cA_{123}\rho \int_0^1 \int_0^1 \int_0^1 \mathbf{a}_{10}^T \mathbf{a}_{10} \xi \, \mathrm{d}\xi \, \mathrm{d}\eta \, \mathrm{d}\zeta \tag{10.77}
$$

where $\mathbf{a}_{10}$ is expressed in terms of the nondimensional coordinates ξ, η, and ζ.

13
Rectangular Hexahedron (Brick) T11: Assumed Displacement Distribution

The node numbering and element forces are shown in Fig. 11.1 for a rectangular hexahedron (brick). At each node point there are three element forces and three corresponding displacements in the x, y, and z directions, respectively.

The assumed displacement field for the hexahedron element (brick) can be taken as

$$u = c_1 + c_2\xi + c_3\eta + c_4\zeta + c_5\xi\eta + c_6\eta\zeta + c_7\zeta\xi$$
$$+ c_8\xi\eta\zeta \tag{11.1}$$

$$v = c_9 + c_{10}\xi + c_{11}\eta + c_{12}\zeta + c_{13}\xi\eta + c_{14}\eta\zeta$$
$$+ c_{15}\zeta\xi + c_{16}\xi\eta\zeta \tag{11.2}$$

$$w = c_{17} + c_{18}\xi + c_{19}\eta + c_{20}\zeta + c_{21}\xi\eta + c_{22}\eta\zeta$$
$$+ c_{23}\zeta\xi + c_{24}\xi\eta\zeta \tag{11.3}$$

where u, v, and w are displacements in the x, y, and z directions within the element. The nondimensional coordinates used here are given by

$$\xi = \frac{x}{a} \tag{11.4}$$

$$\eta = \frac{y}{b} \tag{11.5}$$

$$\zeta = \frac{z}{c} \tag{11.6}$$

Hence the element node displacements **u** can be expressed as

$$
\mathbf{u}=\begin{bmatrix} u_1 \\ u_2 \\ u_3 \\ u_4 \\ u_5 \\ u_6 \\ u_7 \\ u_8 \\ u_9 \\ u_{10} \\ u_{11} \\ u_{12} \\ u_{13} \\ u_{14} \\ u_{15} \\ u_{16} \\ u_{17} \\ u_{18} \\ u_{19} \\ u_{20} \\ u_{21} \\ u_{22} \\ u_{23} \\ u_{24} \end{bmatrix}
=
\begin{bmatrix}
1&0&0&0&0&0&0&0&0&0&0&0 \\
0&0&0&0&0&0&0&0&1&0&0&0 \\
0&0&0&0&0&0&0&0&0&0&0&0 \\
1&1&0&0&0&0&0&0&0&0&0&0 \\
0&0&0&0&0&0&0&0&1&1&0&0 \\
0&0&0&0&0&0&0&0&0&0&0&0 \\
1&1&1&0&1&0&0&0&0&0&0&0 \\
0&0&0&0&0&0&0&0&1&1&1&0 \\
0&0&0&0&0&0&0&0&0&0&0&0 \\
1&0&1&0&0&0&0&0&0&0&0&0 \\
0&0&0&0&0&0&0&0&1&0&1&0 \\
0&0&0&0&0&0&0&0&0&0&0&0 \\
1&0&0&1&0&0&0&0&0&0&0&0 \\
0&0&0&0&0&0&0&0&1&0&0&1 \\
0&0&0&0&0&0&0&0&0&0&0&0 \\
1&1&0&1&0&0&1&0&0&0&0&0 \\
0&0&0&0&0&0&0&0&1&1&0&1 \\
0&0&0&0&0&0&0&0&0&0&0&0 \\
1&1&1&1&1&1&1&1&0&0&0&0 \\
0&0&0&0&0&0&0&0&1&1&1&1 \\
0&0&0&0&0&0&0&0&0&0&0&0 \\
1&0&1&1&0&1&0&0&0&0&0&0 \\
0&0&0&0&0&0&0&0&1&0&1&1 \\
0&0&0&0&0&0&0&0&0&0&0&0
\end{bmatrix}
$$

$$
\begin{bmatrix}
0&0&0&0&0&0&0&0&0&0&0&0 \\
0&0&0&0&0&0&0&0&0&0&0&0 \\
0&0&0&0&1&0&0&0&0&0&0&0 \\
0&0&0&0&0&0&0&0&0&0&0&0 \\
0&0&0&0&0&0&0&0&0&0&0&0 \\
0&0&0&0&1&1&0&0&0&0&0&0 \\
0&0&0&0&0&0&0&0&0&0&0&0 \\
1&0&0&0&0&0&0&0&0&0&0&0 \\
0&0&0&0&1&1&1&0&1&0&0&0 \\
0&0&0&0&0&0&0&0&0&0&0&0 \\
0&0&0&0&0&0&0&0&0&0&0&0 \\
0&0&0&0&1&0&1&0&0&0&0&0 \\
0&0&0&0&0&0&0&0&0&0&0&0 \\
0&0&0&0&0&0&0&0&0&0&0&0 \\
0&0&0&0&1&0&0&1&0&0&0&0 \\
0&0&0&0&0&0&0&0&0&0&0&0 \\
0&0&1&0&0&0&0&0&0&0&0&0 \\
0&0&0&0&1&1&0&1&0&0&1&0 \\
0&0&0&0&0&0&0&0&0&0&0&0 \\
1&1&1&1&0&0&0&0&0&0&0&0 \\
0&0&0&0&1&1&1&1&1&1&1&1 \\
0&0&0&0&0&0&0&0&0&0&0&0 \\
0&1&0&0&0&0&0&0&0&0&0&0 \\
0&0&0&0&1&0&1&1&0&1&0&0
\end{bmatrix}
\begin{bmatrix} c_1 \\ c_2 \\ c_3 \\ c_4 \\ c_5 \\ c_6 \\ c_7 \\ c_8 \\ c_9 \\ c_{10} \\ c_{11} \\ c_{12} \\ c_{13} \\ c_{14} \\ c_{15} \\ c_{16} \\ c_{17} \\ c_{18} \\ c_{19} \\ c_{20} \\ c_{21} \\ c_{22} \\ c_{23} \\ c_{24} \end{bmatrix}
\tag{11.7}
$$

Symbolically the preceding equation can be expressed as

$$\mathbf{u}=\mathbf{C}_{11}\mathbf{c} \tag{11.8}$$

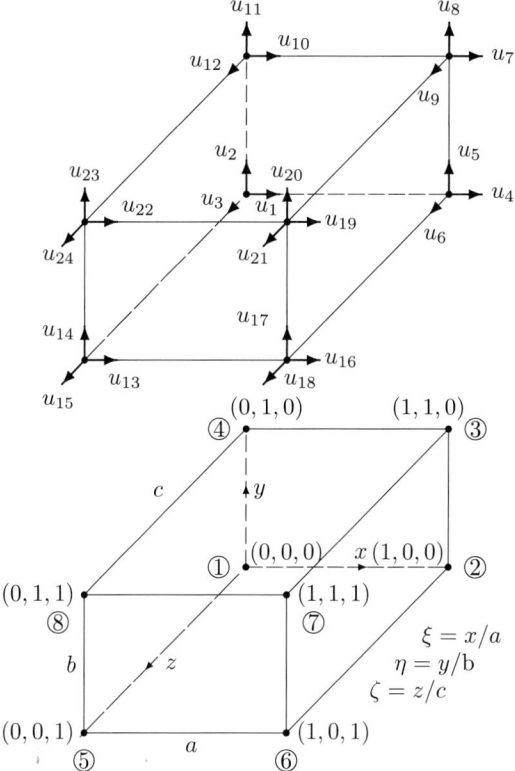

Fig. 11.1   Node displacements and forces for rectangular hexahedron.

Hence,

$$\mathbf{c} = \mathbf{C}_{11}^{-1}\mathbf{u} \qquad (11.9)$$

Noting that $\partial u/\partial x = \partial u/\partial \xi \times \partial \xi \partial x$, etc., the strains $\mathbf{e}$ in the element are then computed from

$$\mathbf{e} = \begin{bmatrix} e_{xx} \\ e_{yy} \\ e_{zz} \\ e_{xy} \\ e_{yz} \\ e_{zx} \end{bmatrix} = \begin{bmatrix} \dfrac{\partial u}{\partial x} \\[2mm] \dfrac{\partial v}{\partial y} \\[2mm] \dfrac{\partial w}{\partial z} \\[2mm] \dfrac{\partial v}{\partial x} + \dfrac{\partial u}{\partial y} \\[2mm] \dfrac{\partial w}{\partial y} + \dfrac{\partial v}{\partial z} \\[2mm] \dfrac{\partial u}{\partial z} + \dfrac{\partial w}{\partial x} \end{bmatrix} = \begin{bmatrix} \dfrac{1}{a}\dfrac{\partial u}{\partial \xi} \\[2mm] \dfrac{1}{b}\dfrac{\partial v}{\partial \eta} \\[2mm] \dfrac{1}{c}\dfrac{\partial w}{\partial \zeta} \\[2mm] \dfrac{1}{a}\dfrac{\partial v}{\partial \xi} + \dfrac{1}{b}\dfrac{\partial u}{\partial \eta} \\[2mm] \dfrac{1}{b}\dfrac{\partial w}{\partial \eta} + \dfrac{1}{c}\dfrac{\partial v}{\partial \zeta} \\[2mm] \dfrac{1}{c}\dfrac{\partial u}{\partial \zeta} + \dfrac{1}{a}\dfrac{\partial w}{\partial \xi} \end{bmatrix}$$

$$= \begin{bmatrix}
0 & 1/a & 0 & 0 & \eta/a & 0 & \zeta/a & \eta\zeta/a & 0 & 0 & 0 & 0 \\
0 & 0 & 0 & 0 & 0 & 0 & 0 & 0 & 0 & 0 & 1/b & 0 \\
0 & 0 & 0 & 0 & 0 & 0 & 0 & 0 & 0 & 0 & 0 & 0 \\
0 & 0 & 1/b & 0 & \xi/b & \zeta/b & 0 & \xi\zeta/b & 0 & 1/a & 0 & 0 \\
0 & 0 & 0 & 0 & 0 & 0 & 0 & 0 & 0 & 0 & 0 & 1/c \\
0 & 0 & 0 & 1/c & 0 & \eta/c & \xi/c & \xi\eta/c & 0 & 0 & 0 & 0
\end{bmatrix}$$

$$\begin{bmatrix}
0 & 0 & 0 & 0 & 0 & 0 & 0 & 0 & 0 & 0 & 0 & 0 \\
\xi/b & \zeta/b & 0 & \zeta\xi/b & 0 & 0 & 0 & 0 & 0 & 0 & 0 & 0 \\
0 & 0 & 0 & 0 & 0 & 0 & 0 & 1/c & 0 & \eta/c & \xi/c & \xi\eta/c \\
\eta/a & 0 & \zeta/a & \eta\zeta/a & 0 & 0 & 0 & 0 & 0 & 0 & 0 & 0 \\
0 & \eta/c & \xi/c & \zeta\xi/c & 0 & 0 & 1/b & 0 & \xi/b & \zeta/b & 0 & \zeta\xi/b \\
0 & 0 & 0 & 0 & 0 & 1/a & 0 & 0 & \eta/a & 0 & \zeta/a & \eta\zeta/a
\end{bmatrix}$$

$$\times \begin{bmatrix} c_1 \\ c_2 \\ \vdots \\ c_{23} \\ c_{24} \end{bmatrix} \qquad (11.10)$$

or symbolically as

$$\mathbf{e} = \mathbf{A}_{11}\mathbf{c} \qquad (11.11)$$

where $\mathbf{A}_{11}$ is the 6×24 matrix in Eq. (11.10). Substituting now the already derived expression for $\mathbf{c}$ in Eq. (11.11),

$$\mathbf{e} = \mathbf{A}_{11}\mathbf{C}_{11}^{-1}\mathbf{u} = \mathbf{b}_{11}\mathbf{u} \qquad (11.12)$$

where

$$\mathbf{b}_{11} = \mathbf{A}_{11}\mathbf{C}_{11}^{-1} \qquad (11.13)$$

Here the matrix $\mathbf{b}_9$ represents a matrix of strains caused by unit displacements at the element nodes. These strains do not satisfy equations of stress equilibrium within the element. Also, all six components of strain are varying linearly throughout the hexahedron element.

The inverse of the matrix $\mathbf{A}_{11}$ is given by

$$
\mathbf{C}_{11}^{-1} =
\left[
\begin{array}{cccccccccccc|cccccccccccc}
1 & 0 & 0 & 0 & 0 & 0 & 0 & 0 & 0 & 0 & 0 & 0 & 0 & 0 & 0 & 0 & 0 & 0 & 0 & 0 & 0 & 0 & 0 & 0 \\
-1 & 0 & 0 & 1 & 0 & 0 & 0 & 0 & 0 & 0 & 0 & 0 & 0 & 0 & 0 & 0 & 0 & 0 & 0 & 0 & 0 & 0 & 0 & 0 \\
-1 & 0 & 0 & 0 & 0 & 0 & 0 & 0 & 0 & 1 & 0 & 0 & 0 & 0 & 0 & 0 & 0 & 0 & 0 & 0 & 0 & 0 & 0 & 0 \\
-1 & 0 & 0 & 0 & 0 & 0 & 0 & 0 & 0 & 0 & 0 & 0 & 1 & 0 & 0 & 0 & 0 & 0 & 0 & 0 & 0 & 0 & 0 & 0 \\
1 & 0 & 0 & -1 & 0 & 0 & 1 & 0 & 0 & -1 & 0 & 0 & 0 & 0 & 0 & 0 & 0 & 0 & 0 & 0 & 0 & 0 & 0 & 0 \\
1 & 0 & 0 & 0 & 0 & 0 & 0 & 0 & 0 & -1 & 0 & 0 & -1 & 0 & 0 & 0 & 0 & 0 & 0 & 0 & 0 & 1 & 0 & 0 \\
1 & 0 & 0 & -1 & 0 & 0 & 0 & 0 & 0 & 0 & 0 & 0 & -1 & 0 & 0 & 1 & 0 & 0 & 0 & 0 & 0 & 0 & 0 & 0 \\
-1 & 0 & 0 & 1 & 0 & 0 & -1 & 0 & 0 & 1 & 0 & 0 & 1 & 0 & 0 & -1 & 0 & 0 & 1 & 0 & 0 & -1 & 0 & 0 \\
0 & 1 & 0 & 0 & 0 & 0 & 0 & 0 & 0 & 0 & 0 & 0 & 0 & 0 & 0 & 0 & 0 & 0 & 0 & 0 & 0 & 0 & 0 & 0 \\
0 & 0 & -1 & 0 & 0 & 1 & 0 & 0 & 0 & 0 & 0 & 0 & 0 & 0 & 0 & 0 & 0 & 0 & 0 & 0 & 0 & 0 & 0 & 0 \\
0 & -1 & 0 & 0 & 0 & 0 & 0 & 0 & 0 & 0 & -1 & 0 & 0 & 0 & 0 & 0 & 0 & 0 & 0 & 0 & 0 & 0 & 0 & 0 \\
0 & -1 & 0 & 0 & 0 & 0 & 0 & 0 & 0 & 0 & 0 & 0 & 0 & 1 & 0 & 0 & 0 & 0 & 0 & 0 & 0 & 0 & 0 & 0 \\
0 & 1 & 0 & 0 & -1 & 0 & 0 & 1 & 0 & 0 & -1 & 0 & 0 & 0 & 0 & 0 & 0 & 0 & 0 & 0 & 0 & 0 & 0 & 0 \\
0 & 1 & 0 & 0 & 0 & 0 & 0 & 0 & 0 & 0 & -1 & 0 & 0 & -1 & 0 & 0 & 0 & 0 & 0 & 0 & 0 & 0 & 1 & 0 \\
0 & 1 & 0 & 0 & -1 & 0 & 0 & 0 & 0 & 0 & 0 & 0 & 0 & -1 & 0 & 0 & 1 & 0 & 0 & 0 & 0 & 0 & 0 & 0 \\
0 & -1 & 0 & 0 & 1 & 0 & 0 & -1 & 0 & 0 & 1 & 0 & 0 & 1 & 0 & 0 & -1 & 0 & 0 & 1 & 0 & 0 & -1 & 0 \\
0 & 0 & 1 & 1 & 0 & 0 & 0 & 0 & 0 & 0 & 0 & 0 & 0 & 0 & 0 & 0 & 0 & 0 & 0 & 0 & 0 & 0 & 0 & 0 \\
0 & 0 & -1 & 0 & 0 & 1 & 0 & 0 & 0 & 0 & 0 & 0 & 0 & 0 & 0 & 0 & 0 & 0 & 0 & 0 & 0 & 0 & 0 & 0 \\
0 & 0 & -1 & 0 & 0 & 0 & 0 & 0 & 0 & 0 & 0 & 1 & 0 & 0 & 0 & 0 & 0 & 0 & 0 & 0 & 0 & 0 & 0 & 0 \\
0 & 0 & -1 & 0 & 0 & 0 & 0 & 0 & 0 & 0 & 0 & 0 & 0 & 0 & 1 & 0 & 0 & 0 & 0 & 0 & 0 & 0 & 0 & 0 \\
0 & 0 & 1 & 0 & 0 & -1 & 0 & 0 & 1 & 0 & 0 & -1 & 0 & 0 & 0 & 0 & 0 & 0 & 0 & 0 & 0 & 0 & 0 & 0 \\
0 & 0 & 1 & 0 & 0 & 0 & 0 & 0 & 0 & 0 & 0 & -1 & 0 & 0 & 1 & 0 & 0 & 0 & 0 & 0 & 0 & 0 & 0 & 1 \\
0 & 0 & 1 & 0 & 0 & -1 & -1 & 0 & 0 & 0 & 0 & 0 & 0 & 0 & 1 & 0 & 0 & 1 & 0 & 0 & 0 & 0 & 0 & 0 \\
0 & 0 & -1 & 0 & 0 & 1 & 0 & 0 & -1 & 0 & 0 & 1 & 0 & 0 & 1 & 0 & 0 & -1 & 0 & 0 & 1 & 0 & 0 & -1 \\
\end{array}
\right]
\tag{11.14}
$$

The stress-strain equations are given by

$$
\sigma = \begin{bmatrix} \sigma_{xx} \\ \sigma_{yy} \\ \sigma_{zz} \\ \sigma_{xy} \\ \sigma_{yz} \\ \sigma_{zx} \end{bmatrix} = \frac{E}{(1+v)(1-2v)}
$$

$$
\times \begin{bmatrix}
(1-v) & v & v & 0 & 0 & 0 \\
v & (1-v) & v & 0 & 0 & 0 \\
v & v & (1-v) & 0 & 0 & 0 \\
0 & 0 & 0 & (1-2v)/2 & 0 & 0 \\
0 & 0 & 0 & 0 & (1-2v)/2 & 0 \\
0 & 0 & 0 & 0 & 0 & (1-2v)/2
\end{bmatrix}
$$

$$
\times \begin{bmatrix} e_{xx} \\ e_{yy} \\ e_{zz} \\ e_{xy} \\ e_{yz} \\ e_{zx} \end{bmatrix} - \frac{\alpha T E}{(1-2v)} \begin{bmatrix} 1 \\ 1 \\ 1 \\ 0 \\ 0 \\ 0 \end{bmatrix} \tag{11.15}
$$

which can be written symbolically as

$$
\sigma = \mathbf{E}_3 \mathbf{e} - \frac{\alpha T E}{(1-2v)} \{1\ 1\ 1\ 0\ 0\ 0\} \tag{11.16}
$$

The expression for the three-dimensional Young's modulus $\mathbf{E}_3$ is given by

$$
\mathbf{E}_3 = \frac{E}{(1+v)(1-2v)}
$$

$$
\times \begin{bmatrix}
(1-v) & v & v & 0 & 0 & 0 \\
v & (1-v) & v & 0 & 0 & 0 \\
v & v & (1-v) & 0 & 0 & 0 \\
0 & 0 & 0 & (1-2v) & 0 & 0 \\
0 & 0 & 0 & 0 & (1-2v) & 0 \\
0 & 0 & 0 & 0 & 0 & (1-2v)
\end{bmatrix} \tag{11.17}
$$

The stiffness matrix for the rectangular hexahedron element is obtained from

$$
\mathbf{k}_{11} = \int_V \mathbf{b}_{11}^T \mathbf{E}_3 \mathbf{b}_{11}\, dV = \iiint \mathbf{b}_{11}^T \mathbf{E}_3 \mathbf{b}_{11} |J(x,y,z)|\, dx\, dy\, dz
$$

$$
= abc \int_0^1 \int_0^1 \int_0^1 \mathbf{b}_{11}^T \mathbf{E}_3 \mathbf{b}_{11}\, d\xi\, d\eta\, d\zeta \tag{11.18}
$$

because

$$
|J(x, y, z)| = \begin{vmatrix} \dfrac{\partial x}{\partial \xi} & \dfrac{\partial x}{\partial \eta} & \dfrac{\partial x}{\partial \zeta} \\[2mm] \dfrac{\partial y}{\partial \xi} & \dfrac{\partial y}{\partial \eta} & \dfrac{\partial y}{\partial \zeta} \\[2mm] \dfrac{\partial z}{\partial \xi} & \dfrac{\partial z}{\partial \eta} & \dfrac{\partial z}{\partial \zeta} \end{vmatrix} = abc \tag{11.19}
$$

Thermal Stiffness

The thermal stiffness is obtained from

$$
\mathbf{h}_{11} = \int_v \mathbf{b}_{11}^T \frac{E}{(1 - 2v)} \begin{bmatrix} -1 \\ -1 \\ -1 \\ 0 \\ 0 \\ 0 \end{bmatrix} dV = \frac{Eabc}{(1 - 2v)} \int_0^1 \int_0^1 \int_0^1 \mathbf{b}_{11} \begin{bmatrix} -1 \\ -1 \\ -1 \\ 0 \\ 0 \\ 0 \end{bmatrix} d\xi\, d\eta\, d\zeta \tag{11.20}
$$

Thermal Load

The thermal load $\mathbf{q}_{11}$ is obtained from

$$
\mathbf{q}_{11} = \alpha T \mathbf{h}_{11} \tag{11.21}
$$

Mass Matrix

Based on the assumed displacement field given by Eqs. (11.1–11.3), the relationship

$$
\begin{bmatrix} u \\ v \\ w \end{bmatrix} = \mathbf{a}_{11}\mathbf{u} \tag{11.22}
$$

can be derived from which $\mathbf{a}_{11}$ is found and then used in

$$
\mathbf{m}_{11} = \rho \int_v \mathbf{a}_{11}^T \mathbf{a}_{11}\, dV = \rho abc \int_0^1 \int_0^1 \int_0^1 \mathbf{a}_{11}^T \mathbf{a}_{11}\, d\xi\, d\eta\, d\zeta \tag{11.23}
$$

It can be shown that

$$\mathbf{m}_{11} = \frac{\rho abc}{216}$$

$$\times \begin{bmatrix}
8 & 0 & 0 & 4 & 0 & 0 & 2 & 0 & 0 & 4 & 0 & 0 & 4 & 0 & 0 & 2 & 0 & 0 & 1 & 0 & 0 & 2 & 0 & 0 \\
0 & 8 & 0 & 0 & 4 & 0 & 0 & 2 & 0 & 0 & 4 & 0 & 0 & 4 & 0 & 0 & 2 & 0 & 0 & 1 & 0 & 0 & 2 & 0 \\
0 & 0 & 8 & 0 & 0 & 4 & 0 & 0 & 2 & 0 & 0 & 4 & 0 & 0 & 4 & 0 & 0 & 2 & 0 & 0 & 1 & 0 & 0 & 2 \\
4 & 0 & 0 & 8 & 0 & 0 & 4 & 0 & 0 & 2 & 0 & 0 & 2 & 0 & 0 & 4 & 0 & 0 & 2 & 0 & 0 & 1 & 0 & 0 \\
0 & 4 & 0 & 0 & 8 & 0 & 0 & 4 & 0 & 0 & 2 & 0 & 0 & 2 & 0 & 0 & 4 & 0 & 0 & 2 & 0 & 0 & 1 & 0 \\
0 & 0 & 4 & 0 & 0 & 8 & 0 & 0 & 4 & 0 & 0 & 2 & 0 & 0 & 2 & 0 & 0 & 4 & 0 & 0 & 2 & 0 & 0 & 1 \\
2 & 0 & 0 & 4 & 0 & 0 & 8 & 0 & 0 & 4 & 0 & 0 & 1 & 0 & 0 & 2 & 0 & 0 & 4 & 0 & 0 & 2 & 0 & 0 \\
0 & 2 & 0 & 0 & 4 & 0 & 0 & 8 & 0 & 0 & 4 & 0 & 0 & 1 & 0 & 0 & 2 & 0 & 0 & 4 & 0 & 0 & 2 & 0 \\
0 & 0 & 2 & 0 & 0 & 4 & 0 & 0 & 8 & 0 & 0 & 4 & 0 & 0 & 1 & 0 & 0 & 2 & 0 & 0 & 4 & 0 & 0 & 2 \\
4 & 0 & 0 & 2 & 0 & 0 & 4 & 0 & 0 & 8 & 0 & 0 & 2 & 0 & 0 & 1 & 0 & 0 & 2 & 0 & 0 & 4 & 0 & 0 \\
0 & 4 & 0 & 0 & 2 & 0 & 0 & 4 & 0 & 0 & 8 & 0 & 0 & 2 & 0 & 0 & 1 & 0 & 0 & 2 & 0 & 0 & 4 & 0 \\
0 & 0 & 4 & 0 & 0 & 2 & 0 & 0 & 4 & 0 & 0 & 8 & 0 & 0 & 2 & 0 & 0 & 1 & 0 & 0 & 2 & 0 & 0 & 4 \\
4 & 0 & 0 & 2 & 0 & 0 & 1 & 0 & 0 & 2 & 0 & 0 & 8 & 0 & 0 & 4 & 0 & 0 & 2 & 0 & 0 & 4 & 0 & 0 \\
0 & 4 & 0 & 0 & 2 & 0 & 0 & 1 & 0 & 0 & 2 & 0 & 0 & 8 & 0 & 0 & 4 & 0 & 0 & 2 & 0 & 0 & 4 & 0 \\
0 & 0 & 4 & 0 & 0 & 2 & 0 & 0 & 1 & 0 & 0 & 2 & 0 & 0 & 8 & 0 & 0 & 4 & 0 & 0 & 2 & 0 & 0 & 4 \\
2 & 0 & 0 & 4 & 0 & 0 & 2 & 0 & 0 & 1 & 0 & 0 & 4 & 0 & 0 & 8 & 0 & 0 & 4 & 0 & 0 & 2 & 0 & 0 \\
0 & 2 & 0 & 0 & 4 & 0 & 0 & 2 & 0 & 0 & 1 & 0 & 0 & 4 & 0 & 0 & 8 & 0 & 0 & 4 & 0 & 0 & 2 & 0 \\
0 & 0 & 2 & 0 & 0 & 4 & 0 & 0 & 2 & 0 & 0 & 1 & 0 & 0 & 4 & 0 & 0 & 8 & 0 & 0 & 4 & 0 & 0 & 2 \\
1 & 0 & 0 & 2 & 0 & 0 & 4 & 0 & 0 & 2 & 0 & 0 & 2 & 0 & 0 & 4 & 0 & 0 & 8 & 0 & 0 & 4 & 0 & 0 \\
0 & 1 & 0 & 0 & 2 & 0 & 0 & 4 & 0 & 0 & 2 & 0 & 0 & 2 & 0 & 0 & 4 & 0 & 0 & 8 & 0 & 0 & 4 & 0 \\
0 & 0 & 1 & 0 & 0 & 2 & 0 & 0 & 4 & 0 & 0 & 2 & 0 & 0 & 2 & 0 & 0 & 4 & 0 & 0 & 8 & 0 & 0 & 4 \\
2 & 0 & 0 & 1 & 0 & 0 & 2 & 0 & 0 & 4 & 0 & 0 & 4 & 0 & 0 & 2 & 0 & 0 & 4 & 0 & 0 & 8 & 0 & 0 \\
0 & 2 & 0 & 0 & 1 & 0 & 0 & 2 & 0 & 0 & 4 & 0 & 0 & 4 & 0 & 0 & 2 & 0 & 0 & 4 & 0 & 0 & 8 & 0 \\
0 & 0 & 2 & 0 & 0 & 1 & 0 & 0 & 2 & 0 & 0 & 4 & 0 & 0 & 4 & 0 & 0 & 2 & 0 & 0 & 4 & 0 & 0 & 8
\end{bmatrix}$$

$$(11.24)$$

14
Rectangular Hexahedron (Brick) T12:
Composite Element with Two Pentahedrons T9

The node numbering for the composite element T12 is shown in Fig. 12.1 for the two pentahedrons representing the hexahedron element T12. This composite element can be used for the upper-bound analysis. At each node point there are three element displacements and corresponding three element forces in the x, y, and z directions, respectively. The computation of the stiffness matrix of the hexahedron T12 involves only the addition of the corresponding stiffness coefficients from the two pentahedrons, each as 18×18 matrices placed into the framework of the larger 24×24 stiffness matrix for the hexahedron. The cutting plane forming the two tetrahedrons is in the plane passing through the node points 1, 3, 7, and 5 as shown in Fig. 12.1. This approach is simpler than the use of the enhanced displacement field vanishing on the boundaries of the element to create the upper-bound element. The node numbering, the node forces, and the corresponding node displacements remain the same as in the T11 element. Furthermore, the mass matrix $\mathbf{m}_{12}$ and the

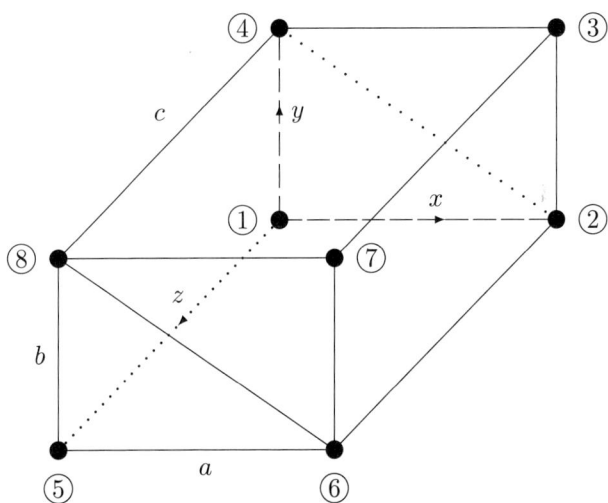

Fig. 12.1 Rectangular hexahedron represented by two prismatic pentahedrons.

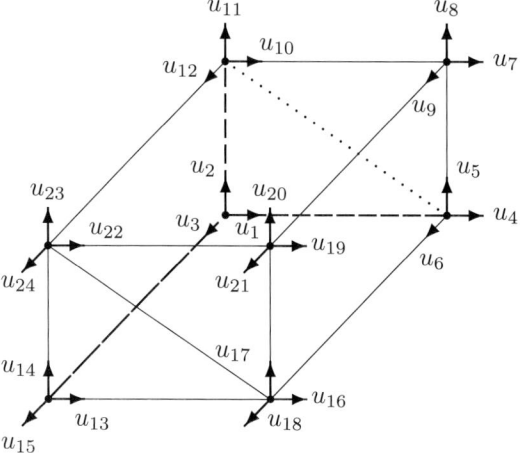

Fig. 12.2   Node displacements and forces for rectangular hexahedron represented by two pentahedrons.

thermal stiffness matrix $\mathbf{h}_{12}$ are assumed to be the same as in the T11 element. The numbering of element displacements and their associated element forces are shown in Fig. 12.2. The cutting plane forming the two tetrahedrons is in the plane passing through the node points 1, 3, 7, and 5 shown in Fig. 12.1.

15
Pin-Jointed Bar and Beam Elements

Some of the elements used currently in the finite element analysis do not have any upper- or lower-bound formulations because their properties derived for the finite element analysis such as stiffness or mass matrices are exact. This is the case of pin-jointed bars and beam elements, which can be referred to as "exact elements." If such elements are needed for the finite element analysis, they can be included with both the upper-bound analysis as well as in the lower-bound analysis.

The pin-jointed bar element has only two axial displacements, one at each end. The beam element, in addition to the axial displacements at each end, has also two bending rotations and one twisting rotation.

Bibliography*

Ainsworth, M., and Oden, J. T., *A Posteriori Error Estimation in Finite Element Analysis*, Wiley, New York, 2000.

Argyris, J. H., Kelsey, S., and Kamel, H., *Matrix Methods of Structural Analysis*, Pergamon, London, 1963.

Brenner, S. C., and Scott, L. R., *The Mathematical Theory of Finite Element Methods*, Springer-Verlag, New York, 2002.

Cook, R. D., et al., *Concepts and Applications of Finite Element Analysis*, Wiley, New York, 1974.

Gupta, K. K., and Meek, J. L., *Finite Element Multidisciplinary Analysis*, 2nd ed., AIAA, Reston, VA, 2000.

Holand, I., and Bell, K., *Finite Element Methods in Stress Analysis*, Tapir, Toronto, 1969.

Hughes, T. J. R., *The Finite Element Method*, 2nd ed., Dover, 2000.

Meek, J. L., *Matrix Structural Analysis*, McGraw-Hill, New York, 1971.

Oden, J. T., *Finite Elements of Nonlinear Continua*, McGraw-Hill, New York, 1972.

Proceedings of the First Conference on Matrix Methods in Structural Mechanics, U.S. Air Force, Wright-Patterson Air Force Base, Dayton, OH, 1965.

Proceedings of the Second Conference on Matrix Methods in Structural Mechanics, U.S. Air Force, Wright-Patterson Air Force Base, Dayton, OH, 1969.

Proceedings of the Third Conference on Matrix Methods in Structural Mechanics, U.S. Air Force, Wright-Patterson, Air Force Base, Dayton, OH, 1973.

Przemieniecki, J. S., *Theory of Matrix Structural Analysis*, McGraw-Hill, New York, 1968; Japanese translation, E. Tuttle Co., Inc., Tokyo, 1971; Chinese translation, Chinese Defense Industry Publishing House, 1974.

Przemieniecki, J. S., *Theory of Matrix Structural Analysis*, McGraw-Hill, New York, 1968.

Robinson, J., *Integrated Theory of Finite Elements*, John Wiley and Sons, London, 1973.

Shames, I. H., and Dym, C. L., *Energy and Finite Element Methods in Structural Mechanics*, McGraw-Hill, New York, 1985.

Tuna, J. J., *Handbook of Structural and Mechanical Matrices*, McGraw-Hill, New York, 1988.

---

*The bibliography contains only some selected textbooks arranged in alphabetical order. At first, the new method of structural analysis was referred to as the matrix methods. It was only later that the new method became known as the finite element method of structural analysis.

Weaver, W., Jr., and Johnston, P., *Structural Dynamics by Finite Elements*, Prentice-Hall, Upper Saddle River, NJ, 1987.

Weaver, W., and Johnston, P. R., *Finite Elements for Structural Analysis*, Prentice-Hall, Upper Saddle River, NJ, 1984.

Young, W. K., and Bang, H., *The Finite Element Method Using MATLAB*, CRC Press, Boca Raton, FL, 2000.

Zienkiewicz, O. C., and Taylor, R. L., *The Finite Element Method for Solid and Structural Mechanics*, 6th ed., Elsevier, New York, 2005.

Index

Supporting Materials

Many of the topics introduced in this book are discussed in more detail in other AIAA publications. For a complete listing of titles in the AIAA Education Series, as well as other AIAA publications, please visit http://www.aiaa.org.